WADSWORTH ANAEROBIC BACTERIOLOGY MANUAL

FIFTH EDITION

Paula Summanen
Ellen Jo Baron
Diane M. Citron
Catherine Strong
Hannah M. Wexler
Sydney M. Finegold

Veterans Administration Wadsworth Medical Center,
and Departments of Medicine, and Microbiology and Immunology,
UCLA School of Medicine, Los Angeles, California

Star
PUBLISHING COMPANY

PUBLISHING COMPANY
P.O. Box 68
Belmont, California 94002

Managing Editor: Stuart Hoffman
Publishers Consultant: Susan G. Kelley, Ph.D.
Project Editor: Ellen Jo Baron, Ph.D.
Cover Design: Douglas Hurd
Typesetting: BookPrep
Photography: Peter Rose

Printed in Singapore


```
               Library of Congress Cataloging-in-Publication Data

Wadsworth anaerobic bacteriology manual. -- 5th ed. / Paula Summanen
   ...[et al.]
        p.    cm.
      Rev. Ed. of: Wadsworth anaerobic bacteriology manual / Vera L.
Sutter, Diane M. Citron, Sydney Finegold. 3rd ed. 1980.
      Includes bibliographical references and index.
      ISBN 0-89863-170-X
      1. Anaerobic bacteria--Identification.  2. Bacteriology--Cultures
and culture media.  3. Pathogenic bacteria--Identification.
I. Summanen, Paula.  II. Sutter, Vera L., 1924- Wadsworth anaerobic
bacteriology manual.  III. Veterans Administration Wadsworth Medical
Center.
      [DNLM: 1. Bacteriological Techniques.  QW 25 W124]
QR67.W33    1993
616'.014--dc20
DNLM/DLC
for Library of Congress                            92-49319
                                                       CIP
```

The authors and the publisher do not imply an
endorsement of the products mentioned in this book.

9 8 7 6 5 4 3 2 1

DEDICATION

We dedicate this fifth edition of the Wadsworth Anaerobic Bacteriology Manual to the memory of Vera L. Sutter, Ph.D. Dr. Sutter was a major force in our laboratory and in anaerobic bacteriology worldwide for many years. She conceived the idea of this manual and was senior author for the first four editions. Vera died on February 15, 1991.

Contents

Preface

Anaerobic bacteria are important pathogens in a wide variety of infections throughout the body. They may be involved in essentially any type of bacterial infection in humans. They play an important role in the most commonly encountered categories of infection—skin and soft tissue, osteomyelitis, pleuropulmonary, intra-abdominal, and female genital tract infection. Most often they are part of a mixed flora, but some infections involve only anaerobes. The severity of anaerobic or mixed anaerobic infections is quite variable—from very low-grade (even inapparent) infection to rapidly progressive infection with significant mortality. Infections involving anaerobes are characterized in particular by a tendency towards suppuration or abscess formation and the production of tissue necrosis. Anaerobic bacteria are not uncommonly resistant to a number of antimicrobial agents; essentially all the mechanisms of resistance encountered in non-anaerobes also operate with the obligate anaerobes.

Anaerobic bacteria may readily be overlooked in infectious processes, particularly when they are present in mixed culture and one or more aerobic or facultative pathogens is grown on conventional culture. Use of relatively crude anaerobic techniques will often permit recovery of the hardier anaerobes such as the *Bacteroides fragilis* group and the *Clostridium perfringens*. This may lull the microbiologist into thinking that his/her techniques for anaerobic culture are entirely adequate since they regularly recover anaerobes. At the present time, however, overlooking anaerobic infections (by the clinician) or anaerobic bacteria (by the microbiologist) is quite uncommon. Indeed, at this time we need to be more concerned about our misplaced confidence that we can readily handle all of these infections because we have antimicrobials available with such excellent activity against anaerobes. This may lead to failure of the clinician to request anaerobic culture or failure of the microbiologist to use optimum anaerobic techniques. This concern is magnified by economic considerations that may dictate eliminating or minimizing anaerobic culture workup. This cavalier approach, when it involves using one of our top drugs with consistent activity against virtually all anaerobes, typically leads (ironically) to increased expense (most of these drugs are expensive) and the real risk of increased resistance to these agents. We are already seeing some resistance to virtually all of our most effective anti-anaerobic drugs in some parts of the world. Even more worrisome is the prospect that these minimal approaches will lead to deterioration of expertise and interest on the part of both microbiologists and clinicians.

This manual is directed primarily at the laboratory worker. Our intent is to emphasize practical approaches to anaerobic bacteriology for clinical laboratories. Certain fastidious anaerobes require very specialized anaerobic techniques but such delicate organisms are seldom involved in infection. We do present some material on more rigorous techniques that are essential for studies of indigenous flora.

We deal virtually exclusively in this manual with organisms encountered in *humans*. Animal strains often have different growth requirements, different taxonomic status, and different patterns of susceptibility to antimicrobial agents.

We would like to express our appreciation to the many individuals, too numerous to name, who have made valuable suggestions or other contributions that have lead to improvements in this monograph since the first edition was published twenty years ago.

Sydney M. Finegold, M.D.

Chapter 1

Introduction To Anaerobic Bacteriology

Anaerobic bacteria are intimately associated with humans, outnumbering aerobes 1,000:1. Some of them behave as opportunistic pathogens when they encounter a permissive environment within the host. A few major syndromes are due to exogenous anaerobes. Special laboratory procedures have been developed for the isolation, identification, and susceptibility testing of this diverse group of bacteria; this manual presents a rational approach to that process.

Many anaerobes grow more slowly than facultative or aerobic bacteria and clinical specimens yielding anaerobic bacteria often contain several organisms and sometimes very complex mixtures. For these reasons, considerable time may elapse before the laboratory is able to provide a final report. Indeed, at times, it may literally take weeks to generate final results from certain specimens with a complex flora. The question then arises: are the bacteriologic data really beneficial to the clinician?

Aside from the time factor, how much information is useful to the clinician? Is the clinician interested in accurate species and subspecies identification or will general identification, with susceptibility data, suffice to permit effective treatment of the patient? Even when the clinician is interested in detailed, accurate bacteriologic data—if only for academic reasons—does cost-benefit analysis warrant providing them? Just how far should the laboratory go in processing anaerobic cultures? These are difficult questions, and the answers vary according to specific circumstances; hard and fast rules cannot be set up.

It is mandatory that the clinician begin therapy empirically, with the assistance of information obtained from gram-stained preparations of appropriate specimens. Essentially all mixed aerobic-anaerobic infections are of endogenous origin; therefore, the clinician or the consultant microbiologist must know the composition of the usual flora at the site where an infection arises and the ways in which that flora may have been modified by various disease states, prior antimicrobial prophylaxis or therapy, and other factors.

In addition, the clinician or microbiologist must be aware of the usual antimicrobial susceptibility patterns of the organisms that might be involved in infection at various sites in the body. This information includes knowledge of the patterns of susceptibility in the particular hospital in which the clinician practices, since there will be variability from one institution to another depending primarily on patterns of antimicrobial use.

Clearly, it is important that the laboratory provide as much information as possible as soon as it can after receipt of the specimen. This is particularly true, of course, in the case of very ill patients. It is the physician's responsibility to call such patients to the attention of the microbiologist. A series of reports at various stages of the laboratory identification process on each individual specimen would provide the clinician with information in optimal fashion. The initial report, at least in the case of very sick patients, should be an interpretation of the Gram stain and any other direct examination of the specimen. The microbiologist must not hesitate to express judgment on likely possibilities, not only from direct examination, but (later) from observation of colonial characteristics and cellular morphology. This presumes, of course, that the microbiologist is well informed and applies rational judgment.

After 18 to 24 hours, examination of aerobic cultures (and certain special anaerobic cultures such as selective media for the *Bacteroides fragilis* group and clostridia) permits the microbiologist to provide a more reliable report. Similarly, examination of routine anaerobic cultures after 48 hours provides even further information, especially when selective and/or differential media are used. A number of rapid diagnostic procedures may be employed to expedite presumptive or definitive identification.

THE INDIGENOUS FLORA

A knowledge of the presence of specific anaerobes as indigenous flora at various sites in the body is important in several ways. Since most anaerobic infections arise in proximity to mucosal surfaces, where anaerobes predominate as indigenous flora, information on which organisms make up the indigenous flora at these sites enables one to anticipate the presence of certain organisms in particular specimens and thus to assist the clinician in choosing the proper drugs for initiating therapy. This information also helps the microbiologist to choose selective and other media that might be particularly useful. Knowledge of the indigenous flora of various regions may also allow one to judge more readily whether a given isolate is significant. For example, *Propionibacterium* in a single blood culture most often represents "contamination" from the patient's skin, particularly if growth does not appear until after several days. Conversely, the presence of *Clostridium septicum* in blood cultures may suggest the portal of entry for the bacteremia and even the patient's underlying condition.

Table 1-1 indicates the incidence of certain anaerobes as indigenous flora at various sites in humans. Further details are provided below. Additional data on anaerobes as indigenous flora are found in Rosebury's classic book [192] and in other references.[18,19,63,75,78,80,168,228] The three major sites of normal colonization of mucosal surfaces are the oral cavity, the intestinal tract, and the female genital tract. Infections relating to these sites of carriage typically involve organisms found normally at those locations as well as other organisms (usually nosocomial pathogens) that colonize patients in the hospital setting. Organisms introduced in the course of surgical procedures or various manipulations may also play a role.

Table 1-1. Incidence of various anaerobes as indigenous flora in humans

	Gram-positive							Gram-negative			
	Clostridium	*Actinomyces*	*Bifidobacterium*	*Eubacterium*	*Lactobacillus*[1]	*Propionibacterium*	Cocci	*B. fragilis* group	*Fusobacterium*	Other GNR	Cocci
Skin	0	0	0	+/-	0	2	1	0	0	0	0
Upper respiratory tract[2]	0	1	0	+/-	0	1	1	0	1	2	1
Mouth	+/-	1	1	1	1	+/-	2	0	2	2	2
Intestine	2	+/-	2	2	1-2	+/-	2	2	1	2	1
Externai genitalia	0	0	0	U	0	U	1	+/-	+/-	1	0
Urethra	+/-	0	0	U	+/-	0	+/-	+/-	+/-	1	U
Vagina	+/-	+/-	+/-	-/-	2	+/-	2	+/-	+/-	1	+/-
Endocervix	+/-	0	0	+/-	1	+/-	2	+/-	+/-	1	+/-

Key:
    U, unknown; O, not found or rare, +/- irregular; 1, usually present; 2, usually present in large numbers

[1] Includes anaerobic, microaerobic, and facultative strains
[2] Includes nasal passages, nasopharynx, oropharynx, and tonsils

The indigenous oral flora consists of various streptococci, particularly those of the viridans group, as well as various nonsporeforming anaerobic bacteria.[192] The nonsporeforming anaerobes include various *Prevotella, Bacteroides, Porphyromonas,* and *Fusobacterium* species (among the gram-negative rods), *Eubacterium* spp., *Lactobacillus, Bifidobacterium, Actinomyces, Peptostreptococcus* and *Veillonella.* The *Bacteroides fragilis* group (most strains of which are β-lactamase producers) is not found often in the indigenous oral flora and takes part in anaerobic pulmonary infections in only about 10% of patients with that type of process. Other gram-negative anaerobic rods may also produce β-lactamase, however, and therefore may be resistant to β-lactam agents such as penicillin. Included among these are the pigmented forms in the genus *Prevotella, Porphyromonas, Prevotella oris* and *P. buccae* (formerly known as *Bacteroides ruminicola*), the *Prevotella oralis* group, the *Bacteroides ureolyticus* group (which includes *Bacteroides gracilis,* an organism resistant to many antibiotics), *Prevotella bivia,* and *Prevotella disiens.* The most common *Fusobacterium* species encountered is *Fusobacterium nucleatum,* but *Fusobacterium necrophorum* is also an important pathogen. Other organisms associated with infections in the oral cavity include *Bacteroides forsythus, Campylobacter rectus,* and *Mitsuokella dentalis. Actinobacillus actinomycetemcomitans,* an aerobic gram-negative bacillus, is important in juvenile periodontitis.

If a patient aspirates in the hospital setting (a relatively common situation), subsequent aspiration pneumonia may involve not only anaerobic bacteria and viridans streptococci, but also such nosocomial pathogens as *Staphylococcus aureus,* various members of the family Enterobacteriaceae, and *Pseudomonas* species. Organisms from the oral cavity, of course, may be involved in various oral and dental infections, head and neck infections, and central nervous system (CNS) infections in addition to pulmonary and pleural infections; they may also be involved in soft tissue infections in "skin poppers" (users who inject illicit drugs subcutaneously), especially if they "lubricate" the needle with saliva.

The flora of the gastrointestinal tract is of interest in connection with intraabdominal infection. Normally the stomach has a sparse flora consisting typically of only about 100 organisms/ml of gastric juice and representing chiefly organisms swallowed from the oral flora.[63] In patients with achlorhydria or a relatively high gastric pH as a result of therapy (e.g., with H_2 antagonists), counts of organisms in the stomach may reach 10^6-10^7/ml or even higher.[63] Certain disease states, such as bleeding or obstructing peptic ulcer, gastric ulcer, or gastric carcinoma, lead to significant colonization of the stomach with a diverse flora that includes members of the Enterobacteriaceae and various anaerobes, even the *B. fragilis* group.[185]

The upper small bowel also has a relatively sparse flora under normal circumstances. In the distal ileum, however, counts of bacteria reach 10^4-10^6/ml, and both coliforms and various anaerobes may be encountered.[63]

The colonic flora is much more profuse and diverse than the small bowel flora. In the distal colon, total counts average 10^{11}-10^{12}/g of feces, with anaerobes outnumbering other organisms by a ratio of 1,000:1.[80] Members of the *B. fragilis* group are dominant in the indigenous flora of the large bowel. Within the *B. fragilis* group, the species that is most prevalent in the indigenous flora is *Bacteroides thetaiotaomicron.* This organism is also the member of the *B. fragilis* group most resistant to antimicrobial agents. In addition to the *B. fragilis* group and other gram-negative anaerobes, the colonic flora includes large numbers of clostridia and anaerobic cocci as well as nonsporeforming, gram-positive anaerobic rods, which are primarily of low pathogenicity. Among the facultative organisms, *Escherichia coli* and various streptococci and enterococci predominate.[80]

In biliary tract infection, *E. coli*—and sometimes similar organisms, such as *Klebsiella* species—and enterococci are the organisms most commonly encountered. *Clostridium perfringens* is not a frequent isolate but may cause devastating disease. In patients in the older age group, those with repeated biliary tract infection related to carcinoma or to intermittent or unrelieved biliary tract obstruction and those with repeated biliary tract surgery, the *B. fragilis* group can also be important.

The flora of the female genital tract

The predominant organisms encountered normally in the female genital tract are various species of *Lactobacillus* and, to a lesser extent, *Peptostreptococcus.* *C. perfringens,* other clostridia, and various gram-negative and gram-positive nonsporeforming rods are encountered irregularly. *P. bivia* and *P. disiens* tend to dominate among the gram-negative rods, but pigmented anaerobic gram-negative bacilli, the *B. fragilis* group, and other *Prevotella* and *Bacteroides* species may be seen as well. *Actinomyces* and *Eubacterium nodatum* are important in women with actinomycotic infection related to intrauterine devices; *Mobiluncus* sp. has been associated with bacterial vaginosis.

Among the nonanaerobes, there are streptococci of groups A and B, other streptococci, and *E. coli* and related organisms. In sexually active women with pelvic inflammatory disease, gonococci, chlamydiae, and *Ureaplasma* may be found; *Mycoplasma hominis,* *Gardnerella vaginalis* and other organisms may be pathogens postpartum.

OPTIMAL ANAEROBIC BACTERIOLOGY WITH RESTRICTED RESOURCES

The desirability of performing optimal anaerobic bacteriology despite limited resources is a real dilemma. Because of the complexity of mixed anaerobic infections and the labor-intensive nature of anaerobic bacteriology, a single culture may cost between $500 and $1,000 or even more. It is important that the microbiologist select approaches that are cost-effective and yet do not compromise good care of the patient.[77]

There are several ways in which costs may be reduced or minimized in the anaerobic bacteriology laboratory: 1) rejection of inappropriate specimens; 2) use of simple, direct tests to provide presumptive evidence that anaerobes are present; 3) use of selective and differential media for rapid isolation and presumptive identification of specific anaerobes; 4) identification of various isclates to a level appropriate for the specimen source, the type of bacteria present, and the patient's status; and 5) use of simple tests for classification of anaerobes into general groups (or definitive identification in the case of some species).

Specimens that may be contaminated with indigenous flora (e.g., throat swabs, expectorated sputum, vaginal swabs) and specimens from relatively minor wounds or lesions that will respond to simple incision and drainage should not be cultured anaerobically. In many anaerobic infections the infecting flora is fairly predictable. If the patient is not very ill, empiric therapy with minimal bacteriologic study is appropriate.

Foul or putrid odor of a specimen is definitive evidence that anaerobic bacteria are part of the infecting flora. Other evidence useful in determining whether anaerobes are present and perhaps in determining which are present include gram-stained preparations with morphology typical of certain anaerobes, fluorescence under ultraviolet light (indicative of pigmented *Prevotella* or *Porphyromonas*), the presence of certain volatile or organic acids on gas chromatography, and (on occasion) the results of DNA probe and fluorescent antibody studies (which will be more valuable when better reagents become available).

Any laboratory engaged in anaerobic cultivation should be able to recover in pure culture all anaerobes present in clinical specimens, to maintain them in a viable state, and to do at least preliminary characterization. This, with identification of certain key organisms such as the *B. fragilis* group and *C. perfringens,* provides the clinician with the data needed to successfully manage patients with anaerobic infections. The above described anaerobes, along with the pigmented *Prevotella* and *Porphyromonas, Fusobacterium nucleatum,* and the anaerobic cocci (these groups are also readily identified) account for approximately two-thirds of all clinically significant infections involving anaerobes.

The use of selective and differential solid media facilitates the work of the bacteriologist.[77,79] Bacteroides bile esculin (BBE) agar [150] is an excellent medium highly selective for the *B. fragilis* group, most strains of which produce large black colonies with blackening of the surrounding area, and for *Bilophila*. Good growth of the *B. fragilis* group may be obtained on BBE medium within 18-24 hours. The appearance of typical colonies of appropriate size on this medium is good presumptive evidence that the *B. fragilis* group is present in a specimen. Other media that are useful include kanamycin-vancomycin laked blood agar, phenylethyl alcohol agar, and cycloserine cefoxitin fructose agar (for *Clostridium difficile*).[85] **Nonselective media should always be used along with selective media.**

As indicated above, the identification of isolates should be tailored to the source of the specimen, the patient's status, and the bacteria present. Close communication (in both directions) between the microbiologist and clinician is required. For very sick patients, an expanded and expedited workup may be of enormous help to the clinician and may indeed be lifesaving. For less significant illnesses, such a detailed workup would merely waste resources.

Whenever possible, simple tests should be used for general—and, to the extent that they permit, definitive—identification of anaerobes.[77] Included in this category would be colonial and microscopic morphology, spot indole, catalase, lipase, lecithinase, the nitrate disk test, the bile disk test, growth stimulation tests, and tests of susceptibility to special potency antibiotic disks (vancomycin, kanamycin, and colistin). Rapid tests, based on the presence of preformed enzymes, may also be useful. It is recognized, of course, that with shortcuts there is always a possibility of error; however, if the clinician and the microbiologist interact and common sense is used, problems at the clinical level should be rare.

In general, the *B. fragilis* group should be identified since it is the most commonly encountered group of anaerobes that may be involved in serious infections and since its members are among the anaerobes most resistant to antimicrobial agents.

Although we recognize that many laboratories may find it difficult to go beyond what has been outlined in the preceding paragraphs and that cost may be a problem, we nonetheless urge full, definitive identification whenever it is possible. As an example, identification to species within the *B. fragilis* group may provide useful information. *B. fragilis* and *B. thetaiotaomicron* are commonly found as pathogens; recovery of *B. merdae* might mean contamination of a specimen with indigenous bowel flora.

Similarly, species identification within the *B. ureolyticus* group is important. *B. gracilis* is much more pathogenic than *B. ureolyticus* and is much more resistant to antimicrobial agents. If an organism initially felt to be *B. fragilis* from a patient with bacteremia of unknown source was subsequently identified as *B.splanchnicus,* this would indicate the bowel as the likely source. If it were *B. fragilis,* the portal of entry might also have been the female genital tract or elsewhere. Identification of a *Clostridium* isolated from the blood as *C. septicum* provides the clinician with a valuable clue, as there is a strong association between bacteremia with this organism and malignancy or other disease of the colon, especially the cecum. Exact identification of an organism isolated from a patient with two or more episodes of infection helps distinguish between recurrence (which may imply tumor, foreign body, or an undrained abscess) and a new infection.

Quantitation of results is of particular importance in anaerobic bacteriology since anaerobic infections are frequently mixed and the relative importance of various organisms in complex mixtures may often be deduced from this type of information. Formal quantitation is not usually necessary; designation as "heavy growth," "few colonies," and so on, is adequate.

It is important that definitive anaerobic bacteriologic studies be carried out in academic centers. One major reason is to avoid deterioration of our present skills; also, such centers are engaged in teaching clinicians and microbiologists as well as pathologists. Moreover, academic centers engage in basic and clinical research that provides information on changes in infection patterns and in susceptibility results; these data help guide empiric therapy in other settings.

CLINICAL BACKGROUND

Incidence of anaerobes in infection

Anaerobes may cause any type of infection in humans. In a number of infections, anaerobic bacteria are the predominant pathogens or are commonly found; these are listed in Table 1-2. When information regarding incidence of anaerobes in these infections is available, it has been indicated.

It must be emphasized, however, that the majority of published studies on anaerobic infections have been retrospective. The bacteriologic methods, particularly the anaerobic methods, were not uniform and, in many instances, not optimal. Therefore, some of these incidence figures are undoubtedly low. Numerous references emphasizing various aspects of anaerobic infections are also listed in Table 1-2 and are recommended for those wishing to read further on this subject.

Table 1-2. Infections Commonly Involving Anaerobes

	% of all cultures yielding anaerobes	Proportion of cultures positive for anaerobes yielding only anaerobes	Reference
Bacteremia	10	4/5	74,248
Bacteremia secondary to tooth extraction	84	21/25	55
Ocular infections	38	10/43	129
Corneal ulcers	7	9/11	191
Central nervous system			
Brain abscess	89	1/2-2/3	105
Subdural empyema	29 (84 cases)		232
Epidural abscess	39 (41 cases)		232
Head and neck			
Chronic sinusitis	52	4/5[1]	83
Acute sinusitis	7		100
Chronic otitis media	56	1/10	33
	59	11/115	7
	33	0	128
Cholesteatoma	92	1/11	116
Neck space infections	100	3/4	14
Wound infection following head and neck surgery	95	0	21
Peritonsillar abscess	94	6/32	35
	76	6/28	82
Dental, oral, facial			
Orofacial, of dental origin	94	4/10	47
Root canal infection	100	18/55	92
	95	13/18	227
Periapical granuloma	86	12/14	118
Periapical abscess	94	16/30	34
Periodontal abscess	100	0/9	184
Dental abscess, endodontic origin	100	8/12	36
Thoracic			

Table 1-2. Infections Commonly Involving Anaerobes

	% of all cultures yielding anaerobes	Proportion of cultures positive for anaerobes yielding only anaerobes	Reference
Aspiration pneumonia	93	1/2[2]	15
	62	1/3	153
	100	1/3	91
Lung abscess	93	1/2-2/3	16
	85	3/4	22
Bronchiectasis	56	0/5	96
	27		199
	17 (18 cases)	3/3	24
Empyema (nonsurgical)	76	29/63	17
Abdominal			
Intra-abdominal	86	1/10	81
infection (general)	90	1/3	173
	81	1/3	233
	94	1/7	93
Appendicitis with peritonitis	92	8/71	23
Liver abscess	52	1/3	195
Other intra-abdominal infection (postsurgery)	93	1/6	95
Wound infection following bowel surgery	60 (33 cases)	5/20	190
Biliary tract	45	0	210
	41	2/117	69
Obstetric-gynecologic			
Miscellaneous types	100	1/3	239
	74	1/3	234
	72		146
Pelvic abscess	88	1/2	2
Vulvovaginal abscess	75	1/4	189
Vaginal cuff abscess	98	1/30	99
Septic abortion, sepsis	69	18/20	46
	67		194
	63		217
Pelvic inflammatory disease	25	1/14	45
	48	1/7	70
Soft tissue and miscellaneous			
Nonclostridial crepitant cellulitis	75	1/12	158
Pilonidal abscess	88 (41 cases)		256

Table 1-2. Infections Commonly Involving Anaerobes

	% of all cultures yielding anaerobes	Proportion of cultures positive for anaerobes yielding only anaerobes	Reference
Bite wound infections:			
Dog bites	30-41		90
Human bites	50-56[3]		90
Diabetic foot ulcers	95	1/20	154
Infected diabetic gangrene (deep tissue culture)	85	1/11	197
Soft tissue abscesses	60	1/4	114
Cutaneous abscesses	62	1/5	167
Decubitus ulcers with bacteremia	63	10/12	44
Osteomyelitis	40	1/10	149
Gas gangrene (clostridial myonecrosis)	100		3
Breast abscess (non-puerperal)	53	5/8	145
	83	18/34	32
	79	5/41	66
Perirectal abscess	77 (74 cases)		256

[1] Twenty-three of 28 cultures (82%) yielding heavy growth of one or more organisms had only anaerobes present.
[2] Aspiration pneumonia occuring in the community rather than in the hospital involves anaerobes to the exclusion of aerobic or facultative forms two-thirds of the time.
[3] Includes clenched fist injuries.

Other studies, not included in Table 1-2, incorporate inaccuracies related to uncertainty about the clinical significance of isolates and to selection of certain types of specimens that would certainly contain elements of the indigenous flora. In this type of study, of course, the incidence of anaerobes may be falsely high.

Table 1-3 details our own recent experience with recovery of anaerobes from clinical specimens.

Clues to anaerobic infection

Certain hints suggest to the microbiologist that a given specimen is likely to contain anaerobic bacteria:

1. Foul odor to specimen.
2. Location of infection in proximity to a mucosal surface.
3. Infections secondary to human or animal bite.
4. Gas in specimen.
5. Previous therapy with aminoglycoside antibiotics (such as gentamicin or amikacin) or other drugs poorly active vs. anaerobes (older quinolones, trimethoprim/sulfame-thoxazole) in the absence of concomitant effective coverage vs. anaerobes.

Table 1-3. Incidence of specific anaerobes in various infections (Wadsworth VA Medical Center experience 1984-1990)

	Blood	CNS[1]	Head & Neck Infections	Dental	Human bites	Animal bites	TTA[2] & pleural	Lung
No. of specimens	161	50	60	12	7	21	90	32
No. of specimens yielding anaerobes	147	19	37	11	3	10	38	11
Bacteroides fragilis	41	2	5	3	0	0	2	0
B. thetaiotaomicron	7	0	3	0	0	0	1	0
Other *B. fragilis* group	14	0	7	0	0	1	4	1
Bilophila wadsworthia	0	0	0	0	0	0	1	0
Prevotella melaninogenica group	0	3	10	5	0	0	18	5
P. intermedia-corporis	3	2	4	10	1	0	9	2
Other pigm. *Prevotella*	1	3	8	8	0	0	11	0
Porphyromonas asaccharolytica	1	0	0	0	0	0	0	0
Porphyromonas sp.	1	2	4	2	0	1	0	1
B. gracilis	1	1	2	0	0	0	1	1
Other *B. ureolyticus* group	2	0	6	6	0	1	4	1
P. oralis group	0	2	8	8	0	0	13	1
Other *Bacteroides/Prevotella* sp.	3	0	5	0	1	0	10	0
Fusobacterium nucleatum	4	3	3	5	1	1	8	2
F. necrophorum	1	0	0	0	0	0	0	0
F. mortiferum/varium	0	0	0	0	0	0	0	0
Other *Fusobacterium* sp.	4	3	5	5	1	2	0	0
Other gram-negative bacilli	2	2	2	3	0	1	5	1
Peptostreptococcus sp.	16	8	20	5	2	3	16	2
Anaerobic *Streptococcus* sp.	3	2	3	6	2	0	7	0
Veillonella sp.	4	0	4	5	0	1	10	1
Other gram-negative cocci	2	0	0	3	1	0	4	0
Clostridium perfringens	32	1	0	0	0	2	1	0
Other *Clostridium* sp.	21	0	0	1	0	1	2	0
Actinomyces sp.	0	7	9	7	1	1	4	0
Bifidobacterium sp.	1	1	0	5	0	0	0	0
Eubacterium sp.	3	2	8	10	1	2	10	0
Lactobacillus sp.	1	1	4	9	0	1	11	6
Propionibacterium acnes	7	11	10	1	1	4	5	3
Propionibacterium sp.	1	1	2	3	0	4	1	1
Other gram-positive bacilli	1	0	1	6	1	1	2	2
Staphylococcus saccharolyticus	1	0	2	0	0	1	1	1

[1] Central nervous system
[2] Transtracheal aspiration

Table 1-3. cont.

	Osteo-[1] myelitis	Foot[1] ulcer	Decubitus ulcer	Perirectal abscess	Miscellaneous soft tissue infections below the waist	Miscellaneous soft tissue infections above the waist
No. of specimens	72	278	15	15	80	33
No. of specimens yielding anaerobes	20	148	8	12	32	9
Bacteroides fragilis	2	17	2	8	9	0
B. thetaiotaomicron	2	1	0	1	4	0
Other *B. fragilis* group	4	15	7	10	19	0
Bilophila wadsworthia	0	0	0	0	0	1
Prevotella melaninogenica group	1	15	0	0	3	2
P. intermedia-corporis	2	6	0	1	4	1
Other pigm. *Prevotella*	3	15	2	3	11	0
Porphyromonas asaccharolytica	0	3	0	0	8	0
Porphyromonas sp.	3	5	0	2	3	0
B. gracilis	1	1	0	2	0	1
Other *B. ureolyticus* group	4	10	0	3	8	0
P. oralis group	0	10	0	1	1	0
Other *Bacteroides/Prevotella* sp.	5	22	1	4	15	3
Fusobacterium nucleatum	0	9	0	1	1	1
F. necrophorum	0	1	0	0	0	0
F. mortiferum/varium	0	0	0	0	0	0
Other *Fusobacterium* sp.	1	5	0	2	1	0
Other gram-negative bacilli	0	3	0	1	7	0
Peptostrepococcus sp.	34	225	11	10	64	9
Anaerobic *Streptococcus* sp.	3	14	2	0	4	1
Veillonella sp.	1	9	0	1	3	0
Other gram-negative cocci	0	2	1	1	2	1
Clostridium perfringens	0	4	0	0	0	0
Other *Clostridium* sp.	0	1	1	1	3	0
Actinomyces sp.	4	7	1	1	7	1
Bifidobacterium sp.	0	0	0	0	1	0
Eubacterium sp.	3	19	1	4	7	3
Lactobacillus sp.	0	12	1	1	3	0
Propionibacterium acnes	1	14	0	1	1	4
Propionibacterium sp.	0	5	1	0	0	3
Other gram-positive bacilli	2	4	0	0	1	0
Staphylococcus saccharolyticus	0	0	0	0	1	0

[1] Designations stated on culture requisition.

Table 1-3. cont.

	Appendiceal tissue or abscess	Peritoneal fluid[1]	Gall bladder	Abdominal, Other
No. of specimens	70	126	20	15
No. of specimens yielding anaerobes	66	74	17	5
Bacteroides fragilis	35	37	3	3
B. thetaiotaomicron	25	34	0	2
Other *B. fragilis* group	60	93	3	8
Bilophila wadsworthia	29	30	0	0
Prevotella melaninogenica group	3	3	0	1
P. intermedia-corporis	7	16	0	1
Other pigm. *Prevotella* sp.	3	5	1	1
Porphyromonas asaccharolytica	2	1	0	0
Porphyromonas sp.	6	3	0	0
B. gracilis	6	9	0	0
Other *B. ureolyticus* group	3	3	0	0
P. oralis group	0	7	0	0
Other *Bacteroides/Prevotella* sp.	29	24	1	2
Fusobacterium nucleatum	10	14	0	0
F. necrophorum	2	2	0	0
F. mortiferum/varium	3	2	0	0
Other *Fusobacterium* sp.	3	7	0	0
Other gram-negative bacilli	8	12	0	0
Peptostrepococcus sp.	34	41	0	3
Anaerobic *Streptococcus* sp.	11	13	1	0
Veillonella sp.	1	3	0	0
Other gram-negative cocci	1	1	0	0
Clostridium perfringens	1	1	0	3
Other *Clostridium* sp.	19	28	3	6
Actinomyces sp.	5	6	0	0
Bifidobacterium sp.	0	0	0	0
Eubacterium sp.	21	29	0	2
Lactobacillus sp.	15	15	0	0
Propionibacterium acnes	5	3	0	1
Propionibacterium sp.	0	2	0	0
Other gram-positive bacilli	12	15	0	1
Staphylococcus saccharolyticus	1	0	0	0

[1] Mostly from appendiceal infections

6. Black discoloration of blood-containing exudates; these exudates may fluoresce red under ultraviolet light (infections involving pigmented gram-negative anaerobes).
7. Presence of ''sulfur granules'' in discharges (actinomycosis).
8. Unique morphology on Gram stain.
9. Failure of organisms seen on Gram stain of original exudate to grow aerobically.
10. Growth in anaerobic zone of fluid media or of agar deeps.
11. Anaerobic growth on media containing 75 to 100 μg/ml of kanamycin, neomycin or paromomycin (or medium also containing vancomycin, in the case of gram-negative anaerobic bacilli) or on other selective media such as BBE, CCFA and RIF (see Appendix A).
12. Characteristic colonies on anaerobic agar plates (for example, *F. nucleatum* and *C. perfringens*).
13. Young colonies of pigmented gram-negative anaerobic rods (growth from anaerobic blood agar plate) that fluoresce red under ultraviolet light.

Taxonomic changes

Tables 1-4 through 1-6 present recent taxonomic changes. Clearly, we must adopt all approved changes in clinical microbiology laboratories. It is prudent to include the previous name in reports for a period of time (as long as one year) to allow clinicians time to adapt. The changes indicated are based on genetic analysis and also take phenotypic characteristics into consideration. Use of the newer terminology allows us to be more precise in our identification. The importance of definitive identification has been discussed in the section entitled ''Optimal Anaerobic Bacteriology in the Face of Restricted Resources.''

PITFALLS

This Manual provides information for collection and processing of clinical material for isolation of anaerobes, from the most basic to very advanced methods. At any level, certain considerations are important. Some of these are listed here.

Common errors made in obtaining and processing clinical specimens:

1. Failure to prepare Gram stain directly from clinical specimen. The Gram stain alerts one to the possible presence of organisms requiring special media or conditions of incubation and indicates that the techniques are inadequate if the organisms seen on the smear fail to grow.
2. Failure to bypass indigenous flora in collecting specimen.
3. Failure to set up anaerobic cultures promptly from clinical specimens or to keep these under anaerobic conditions pending culture.
4. Failure to minimize air exposure during processing.
5. Failure to use a good anaerobic jar. Check for cracks and leaks in plastic lids, O-rings, and vent openings.
6. Failure to use active catalysts when using hydrogen in anaerobic systems. Catalysts must be reactivated after each use by heating to 160°C for 2 hours.

Table 1-4. Recent taxonomic changes

New Nomenclature	Prior Nomenclature	References
Actinomyces georgiae	*Actinomyces D08*	124
A. gerencseriae	*A. israelii serotype II*	124
Anaerorhabdus furcosus	*Bacteroides furcosus*	205,206
Bacteroides caccae	*B. fragilis* "group 3452A"	125
B. forsythus	New species	237
B. galacturonicus	New species	119,120
B. merdae	*B. fragilis* "T4-1" '	125
B. pectinophilus	New species	119,120
*B. salivosus**	New species	156
B. stercoris	*B. fragilis* "subsp. a"	125
*B. tectum**	New species	155
Bilophila wadsworthia	New genus and species	11,12
Campylobacter curvus	*Wolinella curva*	247
C. rectus	*W. recta*	247
Centipeda periodontii	New genus and species	142
Clostridium scindens	New species	176
*Dichelobacter nodosus**	*B. nodosus*	57
Eubacterium yurii	New species	162
*Fibrobacter succinogenes**	*B. succinogenes*	169
*F. intestinalis**	New species	169
Fusobacterium alocis	New species	41
F. periodonticum	New species	213,214
F. sulci	New species	41
F. pseudonecrophorum	New species	211
F. ulcerans	New species	1
Gemella morbillorum	*Streptococcus morbillorum*	136
Lactobacillus oris	New species	73
L. vaginalis	New species	68
Megamonas hypermegas	*B. hypermegas*	203,204
Mitsuokella dentalis	New species	98
M. multiacida	*B. multiacidus*	202,204
Peptostreptococcus hydrogenalis	New species	72
Porphyromonas asaccharolytica	*B. asaccharolyticus*	207
P. endodontalis	*B. endodontalis*	207

** Animal origin*

Table 1-4. Recent taxonomic changes

New Nomenclature	Prior Nomenclature	References
P. gingivalis	*B. gingivalis*	207
Prevotella bivia	*B. bivius*	201
P. buccae	*B. buccae*	201
P. buccalis	*B. buccalis*	201
P. corporis	*B. corporis*	201
P. denticola	*B. denticola*	201
P. disiens	*B. disiens*	201
P. heparinolytica	*B. heparinolyticus*	186,201
P. intermedia	*B. intermedius*	201
P. loescheii	*B. loescheii*	201
P. melaninogenica	*B. melaninogenicus*	201
P. oralis	*B. oralis*	201
P. oris	*B. oris*	201
P. oulorum[1]	*B. oulorum*	201,208
*P. ruminicola**	*B. ruminicola*	201
P. veroralis	*B. veroralis*	201
P. zoogleoformans	*B. zoogleoformans*	201
Propionibacterium propionicum	*Arachnia propionica*	42
*Rikenella microfusus**	*B. microfusus*	52,53
*Ruminobacter amylophilus**	*B. amylophilus*	222,223
*Sebaldella termitidis**	*B. termitidis*	50
Selenomonas artemidis	New species	171
S. dianae	New species	171
S. flueggei	New species	171
S. infelix	New species	171
S. noxia	New species	171
Tissierella praeacuta	*B. praeacutus*	51

** Animal origin*
[1] Minutes of the International Committee on Systematic Bacteriology Subcommittee on Gram-negative Anaerobic Rods meeting on September 13-14, 1990 show a nomenclature correction from *Prevotella oulora* to *P. oulorum.* Int. J. Syst. Bact. 42:590-1, 1991.

Table 1-5. Recent taxonomic changes within the genus *Bacteroides*

New Nomenclature	Prior Nomenclature
B. caccae	*B. fragilis group* "3452A"
B. forsythus	New species
B. galacturonicus	New species
B. merdae	*B. fragilis* "T4-1"
B. pectinophilus	New species
B. salivosus *	New species
B. stercoris	*B. fragilis* "subsp. a"
B. tectum *	New species
Dichelobacter nodosus *	*B. nodosus*
Fibrobacter succinogenes *	*B. succinogenes*
Megamonas hypermegas	*B. hypermegas*
M. multiacida	*B. multiacidus*
Porphyromonas asaccharolytica	*B. asaccharolyticus*
P. endodontalis	*B. endodontalis*
P. gingivalis	*B. gingivalis*
Prevotella bivia	*B. bivius*
P. buccae	*B. buccae*
P. buccalis	*B. buccalis*
P. corporis	*B. corporis*
P. denticola	*B. denticola*
P. disiens	*B. disiens*
P. heparinolytica	*B. heparinolyticus*
P. intermedia	*B. intermedius*
P. loescheii	*B. loescheii*
P. melaninogenica	*B. melaninogenicus*
P. oralis	*B. oralis*
P. oris	*B. oris*
P. oulorum	*B. oulorum*
P. ruminicola *	*B. ruminicola*
P. veroralis	*B. veroralis*
P. zoogleoformans	*B. zoogleoformans*
Rikenella microfusus *	*B. microfusus*
Ruminobacter amylophilus *	*B. amylophilus*
Sebaldella termitidis *	*B. termitidis*
Tissierella praeacuta	*B. praeacutus*

*Animal origin

Table 1-6. Organisms presently included in the genus *Bacteroides*

Bacteroides fragilis group	Other
B. caccae	B. capillosus
B. distasonis	B. cellulosolvens*
B. eggerthii	B. coagulans
B. fragilis	B. forsythus
B. merdae	B. galacturonicus
B. ovatus	B. gracilis
B. stercoris	B. helcogenes*
B. thetaiotaomicron	B. levii*
B. uniformis	B. macacae*
B. vulgatus	B. pectinophilus
	B. polypragmatus*
	B. pneumosintes
	B. putredinis
	B. pyogenes*
	B. salivosus*
	B. splanchnicus
	B. suis*
	B. tectum*
	B. xylanolyticus*

*Animal or environmental origin

7. Failure to use redox indicator or known fastidious anaerobe in anaerobic systems used.
8. Using toxic gas (such as methane from certain municipal sources) in displacement procedure.
9. Failure to include CO_2 in anaerobic gas. Carbon dioxide is essential for growth of some anaerobes.
10. Use of thioglycolate or other liquid medium as the only system for growing anaerobes. A number of anaerobes will not grow in thioglycolate medium, even when it is enriched. Solid media are required to separate various organisms in a mixed culture since rapidly growing organisms may overgrow the anaerobes in liquid medium making recovery difficult.
11. Failure to check liquid medium if growth is recovered from aerobic plates. A Gram stain of the broth often alerts one to the additional presence of anaerobes.
12. Use of inadequate media. Failure to use fresh media.
13. Failure to do quality control on media and reagents for performance.

14. Failure to use supplements in media. Vitamin K_1 and hemin are required by many anaerobic organisms.
15. Failure to use selective media. Some anaerobes may be overgrown by more rapidly growing organisms and overlooked if selective media are not used.
16. Failure to hold cultures for extended periods. Some fastidious organisms, especially if present in small numbers, may require 5-7 days to grow on a plate.
17. Failure to obtain pure culture before performing biochemical tests.
18. Failure to determine whether an organism is a true anaerobe (aerotolerance testing).
19. Failure to determine the Gram reaction correctly (many gram-positive anaerobes overdecolorize and appear gram-negative).
20. Failure to confirm the viability and adequate growth of an organism when doing biochemical tests.
21. Failure to perform confirmation tests when using commercial micromethod identification systems.
22. Inaccurate or incomplete identification.
23. Use of non-standard antimicrobial susceptibility techniques.
24. Failure to use control organisms.
25. Failure to obtain adequate growth in susceptibility tests.
26. Failure to minimize exposure of colonies on plates to air. For some fastidious anaerobes, >15 minutes exposure can cause death.

Commonly encountered errors in identification:

1. A frequent error is inadequate aerotolerance testing of an organism. Many of the fastidious unusual aerobic gram-negative bacilli grow slowly, requiring 48 hours to appear on a plate. Thus the following misidentifications have been made:
   a. *Actinobacillus actinomycetemcomitans* as *Bacteroides* sp.
   b. *Haemophilus aphrophilus* as *B. ureolyticus* group
   c. *Eikenella corrodens* as *B. ureolyticus* group
   d. *Capnocytophaga* sp. as *F. nucleatum*
   e. *Mycoplasma hominis* as *B. pneumosintes*
   f. Nutritionally deficient *Streptococcus* sp. as anaerobic coccus
   g. *H. influenzae* as anaerobic gram-negative bacillus (failure to use chocolate agar for aerotolerance testing)
2. Other common errors due to inadequate testing:
   a. *Clostridium tertium* as *Lactobacillus* sp.
   b. *Bilophila wadsworthia* as *B. ureolyticus*
   c. *B. wadsworthia* as *Desulfomonas* sp.
3. Errors can result from using a non-viable organism to inoculate biochemical or spore tests and then reading the results as valid negatives; always include viability control.
4. Misinterpretation of the Gram reaction may also result in error. An example of this is misidentification of gram-negatively staining *Clostridium* sp., especially *C. clostridioforme*, as *Bacteroides*. These *Clostridium* sp. grow well in 20% bile. Also, coccobacillary gram-negative rods can easily be misinterpreted as anaerobic cocci. Use of special potency antibiotic disks for presumptive identification will differentiate a gram-negative (vancomycin-resistant) organism from gram-positive (vancomycin-sensitive). Gas chromatography will also provide differentiation.

5. Failure to inoculate bile has resulted in misidentification of members of *Prevotella* sp. as *B. fragilis* group organisms and *vice versa*. This may be a problem especially when using rapid identification kits which typically do not include the bile reaction. Several *Prevotella* sp., especially *P. bivia, P. disiens,* and members of the *P. oralis* group, biochemically appear quite similar to *B. fragilis*. Growth in 20% bile or a bile disk susceptibility testing on the purity plate is a key reaction for separating these organisms.

Chapter 2

Specimen Collection and Anaerobic Culture Techniques

SPECIMEN COLLECTION

General Principles

All appropriate specimens collected to exclude indigenous flora and those from normally sterile body sites should be cultured anaerobically. Anaerobic flora are particularly prevalent on mucosal surfaces of the gastrointestinal and genital tracts; specimens collected from these sites should not ordinarily be cultured for anaerobic bacteria. The following specimens are among those likely to be contaminated with indigenous flora and should not be cultured anaerobically under routine circumstances:

Throat swabs
Nasopharyngeal swabs
Gingival or any other internal mouth surface swabs
Expectorated sputum
Sputum obtained by nasotracheal or orotracheal suction
Bronchial washings or other specimens obtained via a bronchoscope, except for those obtained via a protected double lumen catheter or properly done bronchoalveolar lavage
Gastric and small bowel contents
Large bowel contents, except for *C. difficile, C. botulinum, Anaerobiospirillum succiniciproducens,* and other specific etiologic agents
Ileostomy, colostomy effluents
Feces, except as for large bowel contents
Voided or catheterized urine
Vaginal or cervical swabs
Female genital tract cultures collected via the vagina, except for suction curettings or other specimen collected via a double-lumen catheter
Surface swabs from decubitus ulcers, perirectal abscesses, foot ulcers, exposed wounds, eschars, pilonidal sinuses and other sinus tracts
Any material adjacent to a mucous membrane that has not been adequately decontaminated

Table 2-1. Specimen collection for anaerobic microbiology

NOTE: aspirates should be performed after decontamination of intact skin surface with alcohol followed by povidone-iodine.

Site of infectious process	Appropriate specimen	Collection method
Head and neck	Aspirate	Percutaneous needle aspiration
	Tissue biopsy	Surgically obtained
Periodontal	Gingival pocket debris	Sterile paper points into anaerobic transport broth
	Aspirate	Scaler or needle aspiration
Pulmonary	Lung aspirate	Percutaneous needle aspiration of lung
	Tissue biopsy	Surgically obtained tissue biopsy
	Deep bronchial secretions	Transtracheal aspirate or protected bronchial brush
Joint	Joint fluid	Percutaneous needle aspiration
Abdominal	Peritoneal fluid	Percutaneous needle aspiration
	Abscess contents	Aspirate obtained at surgery or under CT or ultrasound guidance (avoiding contamination with bowel contents)
	Bile	Bile obtained at surgery
	Tissue biopsy	Surgically obtained
Female genital tract	Peritoneal fluid	Culdocentesis
	Endometrial material	Endometrial suction or protected collector
	Tissue biopsy (others)	Surgically obtained
Bone	Biopsy	Curettings or scrapings obtained surgically
	Aspirate	Aspirate of deep tissue via uninvolved skin surface
Other soft tissue	Tissue biopsy	Surgically obtained
	Aspirate	Percutaneous needle aspiration
	Tissue	Curettings
Urine	Bladder urine	Suprapubic aspirate

Specimen collection methods

The importance of collecting specimens so as to avoid contamination with indigenous flora cannot be overemphasized. Indigenous anaerobes are often present in such large numbers (10^9-10^{12}) that even minimal contamination with indigenous flora will yield very misleading results and lead to much wasted effort by the laboratory. Recommended collection methods to avoid indigenous flora are listed in Table 2-1.

Methods for collection of appropriate specimens from selected infectious processes, particularly for those sites for which special techniques are often necessary, are described in the following section. There is no specimen for which a swab is the ideal collection method; only when no other options are available and resulting information is judged to be important clinically should a swab be accepted for anaerobic culture.

Any aspirated material from a normally sterile body site is acceptable; skin preparation should be the same as that used for venipuncture for blood culture: vigorous scrubbing with 70% alcohol from the center of the site outward in widening concentric circles to remove surface dirt and oils, followed by an iodine preparation (applied in widening concentric circles from the center outward and allowed to remain wet for 30 sec [tincture of iodine] or 1 minute [aqueous iodophor]). The iodine preparation should be removed with additional alcohol after the procedure, as many people are sensitive to iodine.

Abscesses. As for all loculated fluid collections, material should be aspirated via needle and syringe through disinfected, uninvolved and intact tissue. If the abscess fluid is exposed at surgery and a needle is contraindicated due to small volume of material (less than 0.2 ml) or proximity to critical tissues (such as may occur with brain abscess), fluid may be aspirated through a flexible plastic catheter or directly into a sterile syringe without a needle. A fresh, sterile needle should be used to inject the material from the syringe into an anaerobic transport vial, so as not to carry over contamination from the needle used to collect the specimen.

Sinus tracts or deep, draining wounds (Figure 2-1). The skin surface surrounding the sinus tract should be cleaned thoroughly, first with 70% alcohol and then with an iodophor, which is allowed to remain wet on the skin for a minimum of 30 sec (tincture) or 1 minute (aqueous iodophor). Both disinfectants should be applied in a circular motion, moving outward in concentric circles to a radius of approximately 2 cm beyond the sinus tract. Iodine preparation should also be instilled into the opening of the sinus tract at the same time to disinfect the proximal area. The iodine is removed by aspirating, and small curettings of material from deep within the tract are obtained. Surface curettings should be discarded and deeper curettings should then be obtained for culture.

Alternatively, if pus is draining, aspiration of material from within the tract via a flexible plastic catheter and syringe usually provides adequate material for culture. Cultures prepared from a swab inserted into the exterior opening are not as likely to correctly predict the etiologic agent of the infection; secondary colonizers contaminating the wound will be recovered in greater numbers.

Perioral or gingival abscesses.[34,36,47,184,261] The flora present in oral secretions is commonly found in perioral abscesses; therefore it is essential to avoid contamination during the collection process. Swabs from oral sites are **never** acceptable for anaerobic culture. Cultures of noninfected oral sites, such as subgingival plaque, should be attempted for research purposes only and only by laboratories experienced in the specialized techniques required for such studies. (See Appendix A.)

Aspirated abscess material is the most desirable specimen. Aspiration and drainage via the external skin surface provides the preferred specimen. If the oral cavity mucous membranes must be entered, the site should be isolated with cotton rolls, dried, and swabbed vigorously with povidone-iodine, which is allowed to remain on the site for 1 minute before the needle is inserted. Abscess contents are then aspirated. The needle is

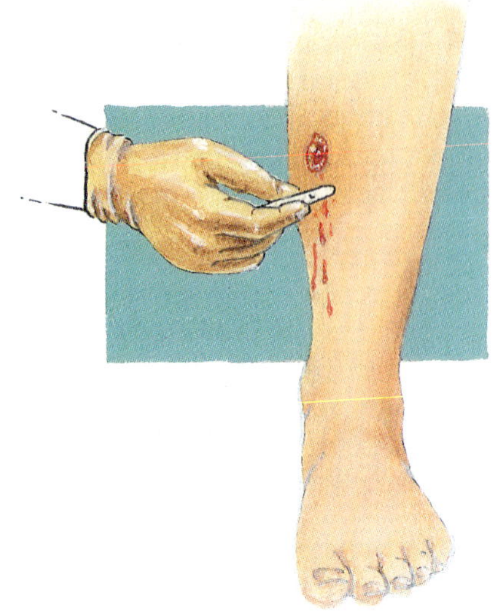

STEP 1: Thoroughly cleanse skin surface surrounding wound with povidone-iodine. Allow to remain wet for 1 min. Wash with ethanol to remove iodine. Allow ethanol to dry.

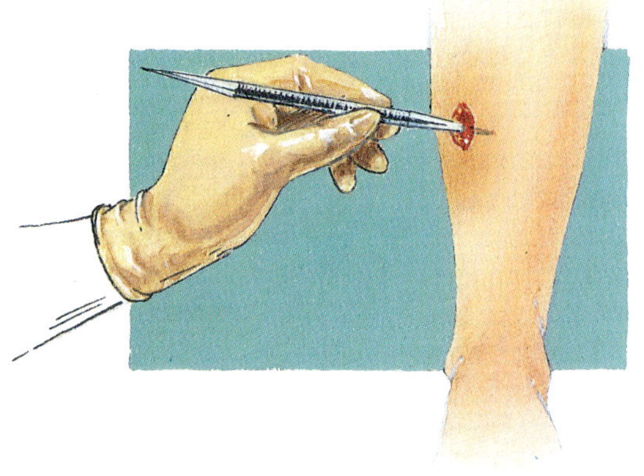

STEP 2: Obtain curetting from deep interior of wound or sinus tract.

NOTE: Swabs are not recommended because they are unlikely to yield clinically significant anaerobes.

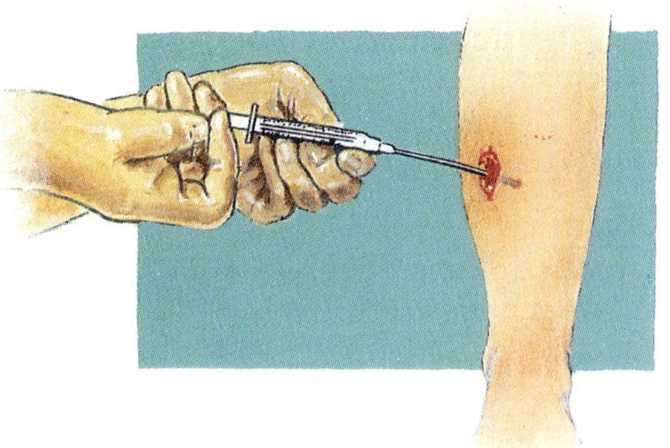

ALTERNATIVE METHOD: Obtain exudate from deep interior of wound using a plastic catheter and syringe.

Figure 2-1. Aspiration of draining sinus tract or deep wound. (From Finegold, Baron, and Wexler, *A Clinical Guide to Anaerobic Infections,* 1992. Courtesy Princeton Scientific Productions, Inc. and Star Publishing Co.)

replaced with a fresh sterile needle, and after pushing the air out of the new needle with some of the specimen, the specimen is expelled into an anaerobic transport vial.

If periodontal areas are being sampled directly, a suitable reduced transport broth should be available for direct inoculation. The area to be sampled should be isolated with cotton rolls and dried with sterile cotton swabs; 70% ethanol may be used to further disinfect and dry the surface. Supragingival plaque must be removed, either with sterile toothpicks or by using a sterile scaler. To obtain the sample, either a sterile scaler (nickel-plated Morse 00 is recommended) or sterile Gracey curette can be used to gently scrape material from the depth of the sulcus.

Sterile paper points are an alternative; insert three fine paper points to the depth of the pocket and leave in place for 10 seconds. Whatever the method of obtaining the specimen, it must be placed immediately into prereduced diluent (chopped meat broth, thioglycolate, Ringer's, or yeast extract). Because of the fastidious and oxygen-sensitive nature of many oral anaerobes, prompt processing (preferably in an anaerobic chamber) is essential to obtaining clinically relevant results.

Paranasal sinus secretions.[83,100,130] Material collected via the nares or upper respiratory tract secretions submitted on swabs are not appropriate for anaerobic culture. Maxillary sinus secretions are obtained by first spraying the nasal cavity below the anterior portion of the inferior turbinate with 10% xylocaine and then swabbing the area to be punctured with a cotton swab soaked in 2% tetracaine to anesthetize and disinfect the area. After waiting 15-20 minutes, material from the maxillary antrum is aspirated from the sinus by a properly trained physician using a sterile 2 mm diameter puncture needle attached to a 20 ml syringe. If no material is obtained initially, the needle can be left in place and the syringe filled with 1-2 ml sterile, reduced diluent, which is injected into the sinus and aspirated.

After the specimen is obtained, a fresh needle is substituted, air is pushed out of the new needle with sample, and the sample is injected into an anaerobic transport vial. Material from other sinuses is collected only during the course of surgical procedures. Endoscopically obtained secretions may also be satisfactory.

Superficial ulcers (i.e., decubitus ulcers, foot ulcers in diabetics)(Figure 2-2). Although these infections commonly involve anaerobes, they are also predisposed to contamination with fecal or other organisms. Thus, collection of material from below the surface is essential. Several strategies for removing contaminated surface tissue are available, including surface debridement with pulsed or gravity-feed iodophor followed by sterile saline, surface debridement by application of wet-to-dry gauze dressing changes over several days, or direct manual debridement with sterile gauze. Following each of these procedures, superficial materials or curettings are discarded, and material from the base of the ulcer is collected by curette or vigorous swabbing; the specimen should be placed immediately into an anaerobic transport tube.

Alternatively, collection of purulent material from under skin flaps or from deep pockets using a needle and syringe inserted through disinfected, uninvolved skin may be used. Each of these methods carries some risk of spreading the infection, either into contiguous tissue or into the bloodstream.

Submarginal irrigation-aspiration has been suggested as a less traumatic alternative method for obtaining anaerobic specimens from draining decubitus ulcers (and probably for other ulcers).[67]

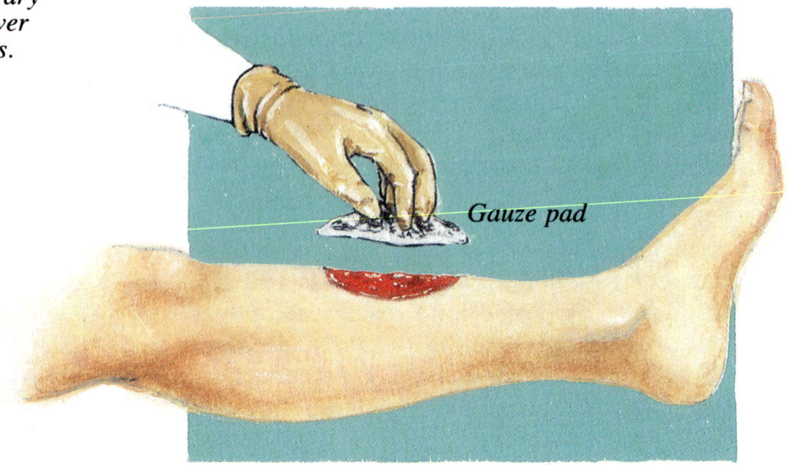

STEP 1: Prepare wound by cleansing with wet-to-dry dressings over several days.

STEP 2: Flush wound with 5 liters of povidone-iodine/sterile saline (50:50 solution) using gravity feed or pulsed jet.

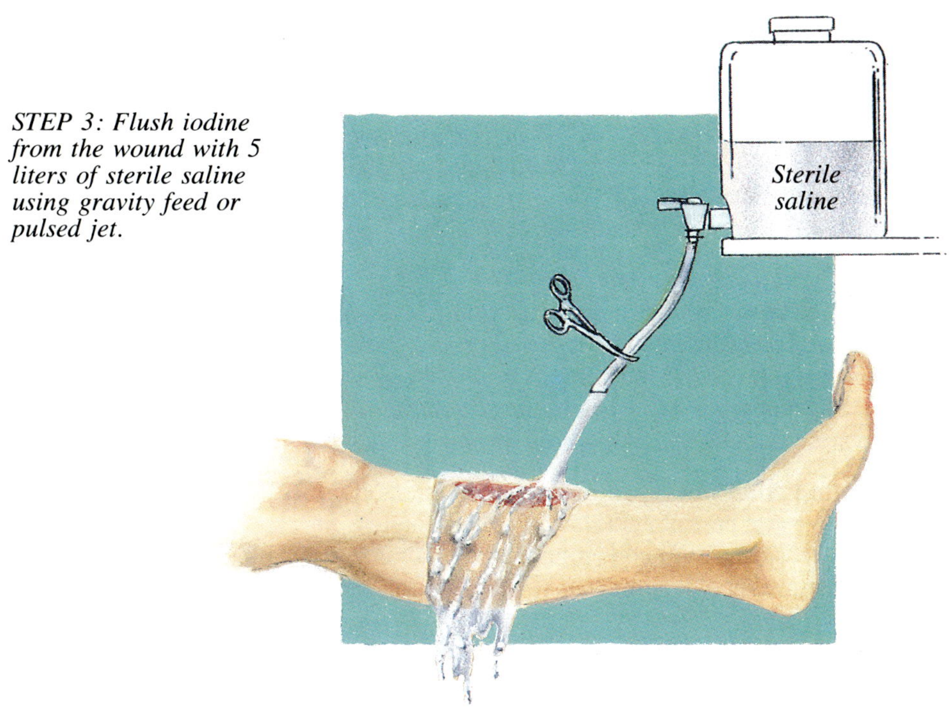

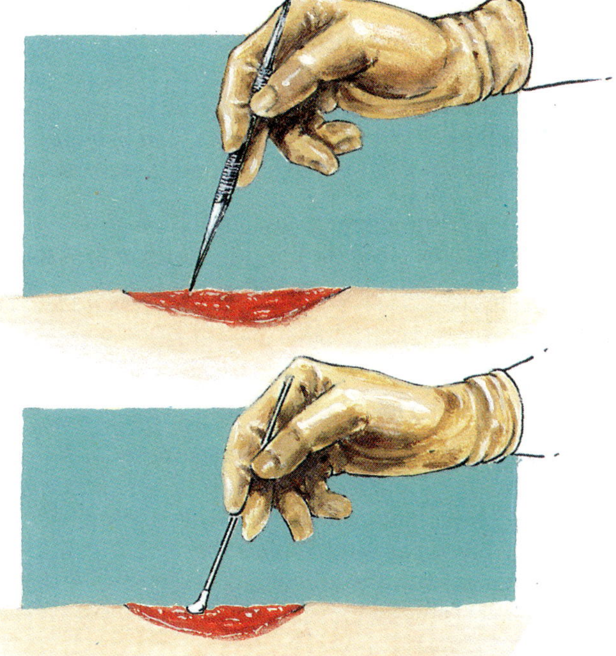

Figure 2-2. Obtaining specimen from contaminated surface wound or ulcer when aspiration is not possible. (From Finegold, Baron, and Wexler, *A Clinical Guide to Anaerobic Infections,* 1992. Courtesy Princeton Scientific Productions, Inc. and Star Publishing Co.)

Respiratory tract secretions.   Most respiratory tract specimens collected via the oral cavity are not suitable for anaerobic culture because of the substantial indigenous flora contamination encountered in specimen collection. Specimens in this category include: expectorated sputum, aspirated sputum, induced sputum, bronchial washings, standard bronchoalveolar lavages, and bronchial brushings obtained without a protected double-lumen catheter. Respiratory specimens acceptable for anaerobic culture include: lung tissue, percutaneous lung or transtracheal aspirates, bronchial brushings collected via a double-lumen protected catheter, "protected" bronchoalveolar lavage, and thoracentesis fluid.

Protected bronchial brush specimens must be placed immediately into reduced anaerobic transport fluid (obtained from the laboratory before beginning the procedure) to serve as the initial diluent and to maintain anaerobic conditions (see procedure in Appendix A). Other specimens obtained with a needle and syringe should be injected into an anaerobic transport vial after a fresh needle has been substituted.

Female genital tract specimens.   Specimens obtained via the vagina are contaminated with large numbers of the same organisms likely to be involved in the pathologic process; therefore, specialized collection methods for anaerobic culture are required. Although culdocentesis yields excellent specimens, this procedure is only used occasionally at present.

After the cervical os has been adequately disinfected by wiping off excess mucus and then swabbing with povidone-iodine, a protected sampling device may be inserted for collection of intrauterine material. One sampling device is thought to adequately recover organisms involved in upper genital tract infections (most useful after the cervical os has been dilated): the endometrial suction curette (Pipelle). The AccuCulShure sintered plastic collection device may also prove to be useful. Both devices include a double-lumen collector and a self-contained transport system. The Pipelle uses suction to obtain cellular material from the uterine wall; this is the most desirable specimen (see procedure in Appendix A). Pathogens other than bacteria, including *Chlamydia,* mycoplasmas, and viruses should also be sought in these specimens.

Specimens from upper vaginal tract infections, such as tubo-ovarian abscess material, should be collected at the time of surgery in the same manner as described for abscesses or tissue.

Swabs of vaginal discharge are not appropriate for anaerobic culture, although a Gram stain can be used to diagnose bacterial vaginosis, a process whose etiology involves mixed anaerobic bacteria.[141]

Specimens collected at time of surgery.   Any tissue obtained at surgery should be appropriate for anaerobic culture. The surgeon should place a small portion of tissue (5 mm^3) representing the infected site (necrotic or gangrenous tissue, abscess wall, etc.) directly into an anaerobic transport tube with a screwcap and an agar plug in its base for maintaining moisture and reducing oxygen. A large piece of tissue (greater than 1 cubic centimeter) may be transported in a sterile petri dish or urine cup if laboratory processing occurs within 1-2 hours.

Other specimens.   Certain other types of specimens may be suitable for anaerobic culture if they are collected to exclude air and contamination, such as urine obtained via a suprapubic bladder tap. Other specimens, although not suitable for routine anaerobic culture, may be processed to determine the presence of a particular species or group (e.g., stool culture for *Clostridium difficile* only, vaginal discharge material for *Bacteroides fragilis* group organisms, or cervical discharge for *Actinomyces* species or *E. nodatum* from a woman with an intrauterine device). In such cases, selective media must be used.

SPECIMEN TRANSPORT

Specimens must be protected from the deleterious effects of oxygen until they can be cultured. In a proper anaerobic transport medium, anaerobic bacteria may survive for up to several days, depending on the nature of the specimen. Purulent specimens contain numerous reducing compounds and are more protective of anaerobic viability than are clear fluids. Anaerobes survive well in pieces of tissue, especially larger ones. Certain specimens, such as stool and urine, contain enzymes that degrade bacterial components; these specimens must be plated as soon as possible. Specimens should be transported and held at room temperature; incubator temperatures will cause differential bacterial over-growth or loss of some strains and cold temperatures will allow increased oxygen diffusion.

Do not transport aspirated material in the syringe! Not only is the transport of an unsheathed (even corked) needle extremely unsafe, but even without a needle attached, the syringe itself may easily be jostled or pushed, which will eject the material and create a hazard. Aerosols may be created, nearby people and the environment may become contaminated, or the specimen may be lost. Several companies produce vials containing reducing substances in an anaerobic environment, capped with a Hungate-style cap or with a rubber septum, through which the specimen can be injected (Figure 2-3). These systems maintain relative numbers and viability of anaerobes very well.[31]

Body fluids may be transported in sterile tubes if anaerobic transport vials are not available. Particularly if they are purulent and >1.0 ml in volume, they will maintain viability adequately for several hours. Clear fluids should be placed into a small container (tube, vial) with as little airspace above the fluid level as possible. The tubes should be kept upright to avoid mixing with air and maintained at room temperature.

Anaerobic transport vials usually contain modified Cary-Blair or other medium containing agar, reducing substances to scavenge excess oxygen, and an oxygen tension indicator (usually resazurin). Tissue pieces and curettings should be placed into either an anaerobic transport vial with a screwtop Hungate cap (Anaerobe Systems) or into a small, sterile tube that is immediately placed under anaerobic conditions. Sealable plastic bags and components to generate an anaerobic atmosphere inside the bag are available commercially (BBL, Becton-Dickinson, Merck); these are very useful for transporting tubes of specimens (with caps loose until the atmosphere is anaerobic) or tissues in petri dishes. Very large pieces of tissue may be transported in a sterile urine cup if they are to be processed within a reasonable time after collection. The interior of such specimens will probably remain reduced over some time (up to several hours at least) if they remain moist.

For swabs (the least desirable specimen), anaerobic transport medium in soft agar deeps is available from BBL and Anaerobe Systems (Figure 2-4). The swab should be placed deep into the agar butt, the swab stick broken off high where held, and the cap replaced quickly.

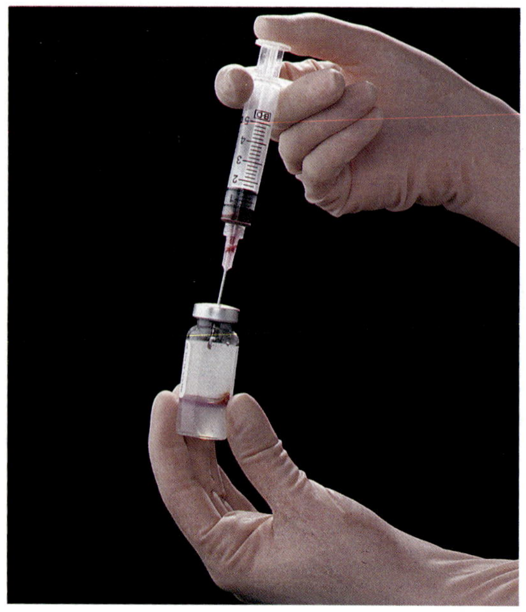

Figure 2-3. Anaerobic transport vial for liquid specimens. Inject specimen without introducing air into the vial. (Caution: inject slowly to ensure that specimen remains on top of agar) (From Finegold, Baron, and Wexler, *A Clinical Guide to Anaerobic Infections,* 1992. Courtesy Princeton Scientific Productions, Inc. and Star Publishing Co.)

STEP 1. Insert swab deep into agar (contains reducing substances to remove residual oxygen).
STEP 2. Break off contaminated end of swab shaft.
STEP 3. Immediately replace cap.

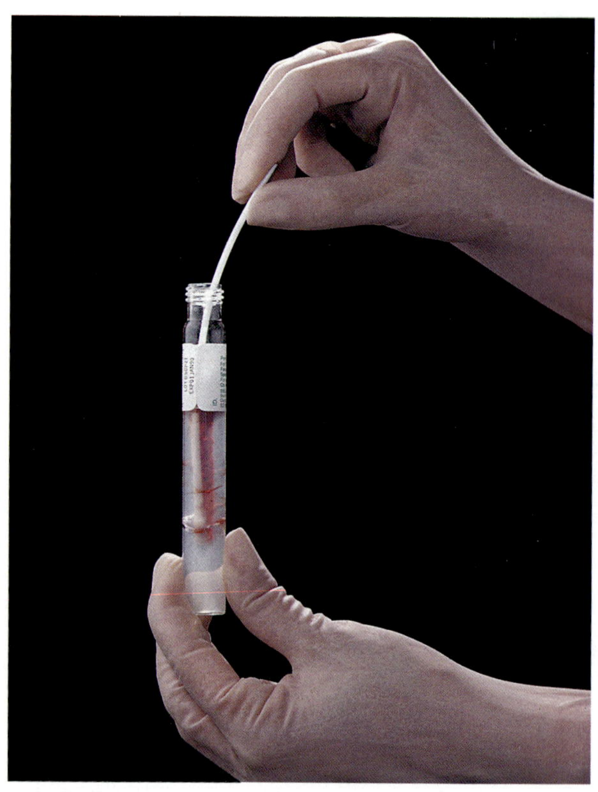

Figure 2-4. Anaerobic transport tube with agar deep for transporting swabs when no other specimen can be obtained. (From Finegold, Baron, and Wexler, *A Clinical Guide to Anaerobic Infections,* 1992. Courtesy Princeton Scientific Productions, Inc. and Star Publishing Co.)

ANAEROBIC CULTURE TECHNIQUES

Anaerobic methods include the use of anaerobic jars, plastic anaerobic bags, the roll tube method of Hungate [113] (and modifications), and the anaerobic chamber or glove box. The roll tube procedures are rather time-consuming, require complex equipment, and utilize PRAS media for identification. Anaerobic chambers, especially the glove-free models, utilize standard methods and are convenient and cost-effective when larger numbers of specimens are processed. They are essential for work on anaerobes of the indigenous flora as these organisms are more oxygen-sensitive than clinical isolates. Two comparative studies have shown that when clinical specimens are collected, transported, and processed properly, recovery of clinically significant anaerobes is as good with anaerobic jars as with the more complex methods. [135,193]

We do not recommend the large diameter Gas Pak-150 jar for incubation of primary plates or the large free-standing anaerobic incubators for routine work, as they do not minimize oxygen exposure. Small incubators inside anaerobic chambers are particularly useful, as the plates can be inspected at 24 hours or earlier for rapidly growing anaerobes such as the *B. fragilis* group and *Clostridium* sp. if necessary. In anaerobic chambers which require the use of desiccants, store the plates in a gas-permeable plastic bag while incubating to avoid drying out the media.

Use of a liquid medium as the only anaerobic culture technique is not satisfactory except in the case of blood cultures. Anaerobic infections usually contain multiple organisms and it might be difficult to isolate the individual components of the mixture from a liquid medium. Facultative organisms may overgrow the anaerobes in broth, making quantitation not possible. Accordingly, liquid media should be used only as an adjunct to other anaerobic culture methods.

Blood culture techniques

Liquid media. There are a number of commercially available media that appear to be satisfactory for recovery of anaerobic bacteria. [248] Some contain sodium polyanethol sulfonate (SPS), which has been reported to enhance the recovery of some anaerobes but may be inhibitory to *Peptostreptococcus anaerobius*. This inhibitory effect can be overcome by addition of 1.2% gelatin. Some of the media available in unvented blood culture bottles, with or without SPS, and prepared under vacuum with CO_2 added, are tryptic soy broth (Difco), thiol broth (Difco), Columbia broth (Difco), trypticase soy broth (Becton Dickinson), thioglycolate medium 135C (BD), and Septi-Chek (BD).

Automated systems. The automated systems that will be discussed are the BACTEC (Becton Dickinson) 460, the BACTEC NR-660, and the BacT/Alert (Organon Teknika Corp.). The BACTEC 460 system measures the uptake of radio-labeled substrates by the organism(s) growing in the bottle. This consumption releases radioactive CO_2 which is measured by the instrument. When the CO_2 concentration reaches a specific threshold value the culture is considered positive. Problems cited with the 460 system include costs and potential hazards of storage, monitoring and disposal of radioactive material. The NR-660 system applies the same principle as the 460 but no radioisotopes are used. An infrared spectrometer is used to measure the CO_2 concentration in the headspace of the blood culture bottle. Studies performed indicate no significant difference in the recovery of anaerobic bacteria or in the mean time for detection of anaerobes between the 460 and

Table 2-2. Comparison of culture methods for isolation of anaerobes from clinical specimens

FACTORS	PRAS tubes	Anaerobic chamber	Anaerobic jar	Anaerobic Bag Technique
Initial cost	Moderate	High	Moderate	Low
Continuing cost	Moderate	Moderate	Moderate	Moderate
Time factor	Moderate	Moderate	Low	Low
Space required	Moderate	Moderate-Large	Moderate	Small
Recovery of significant anaerobes	Satisfactory	Satisfactory	Satisfactory	Satisfactory
Suitability for specimen processing*	Not suitable	Allows processing without exposure to oxygen	Not suitable	Not suitable
Principal advantages	Tubes can be inspected at any time without disturbing anaerobic conditions	Plates can be inspected at any time under anaerobic conditions; conventional isolation procedures are employed	Uncomplicated and convenient; conventional techniques employed	Plates can be inspected at any time without disturbing anaerobic conditions; uncomplicated and convenient; conventional techniques employed
Principal disadvantages	Technique requires more training time; cumbersome to isolate and purify strains; colonial morphology frequently not distinctive	Space requirement; high initial cost	Possibility of prolonged exposure of plates to oxygen during inspection and subculture	Holds only two plates; possibility of prolonged exposure of plate to oxygen during subculture

*Tissue grinding, specimen homogenization with a blender

NR-660 system.[131] While both systems have had problems with false positive results due to carryover, the reliability, ease of operation and lack of radioisotopes are significant advantages of the NR-660 system.

Another automated system designed to detect anaerobic growth in blood culture bottles is the BacT/Alert. It differs from other systems currently available by requiring minimal sample invasion and allowing virtually continuous monitoring of all bottles. A semi-permeable membrane and a colored CO_2 sensor are located at the bottom of each culture bottle. If bacteria are present they generate CO_2 which diffuses through the membrane and reacts with the sensor to produce hydrogen ions. The sensor registers the subsequent drop in pH by a gradual color change from blue-green to yellow. This, in turn, alters the amount and type of light reflected from the sensor. The sensor is illuminated by an LED light source. Reflected light is converted by a photodiode to a signal that is proportional to the concentration of CO_2 in the bottle. The instrument is highly automated; once a bottle is entered into the system, it need not be manipulated until it is removed. Positivity is dependent on either changes in CO_2 level, total level of CO_2, or initial CO_2 level. Unless the bottle is read as positive, it is not handled by a technologist. Although preliminary studies are promising, no long-term data on recovery of anaerobes have been generated yet.[240,263]

Standard blood culture processing.   After properly cleaning and disinfecting the skin, draw blood from the patient and inoculate it immediately into the medium at a ratio of 1 ml of blood to 10 ml of medium. It is advisable to inoculate two bottles (containing different media compositions), venting one for maximal recovery of strict aerobes (these should also be checked for anaerobes) and leaving the other unvented for recovery of anaerobes. When the bottle is inoculated with the patient's sample, air may be inadvertently introduced into the system (although, for most systems, it shouldn't be if the proper precautions are taken); therefore, to avoid further aeration, do not shake the anaerobic bottles. Examine cultures daily for turbidity, colonies, hemolysis, or gas. Gram-stained smears and "blind" aerobic subcultures should be made routinely after the first 6 to 12 hours of incubation. All bottles showing macroscopic growth should be subcultured anaerobically on a blood agar plate.[180,187] Negative culture bottles should be held for 7 days.

Anaerobic incubation methods (Table 2-2)

Jar techniques.   Anaerobic jars are used primarily with plated media, both primary plates and subcultures (Figure 2-5). In order to obtain reliable results from any anaerobic jar there must be adequate replacement of the oxygenated environment with an anaerobic atmosphere. This is achieved by introducing a gas mixture containing H_2 into the jar. In the presence of a catalyst, the oxygen combines with the H_2 to produce water, thus establishing anaerobiosis. A catalyst composed of palladium-coated alumina pellets (Coy Laboratory Products, Anaerobe Systems, and Engelhard Industries) held in a wire mesh basket is preferred, as it is convenient to use and there is no explosion hazard with this "cold" catalyst. The catalysts must be regenerated after each use, as excess moisture and H_2S will inactivate the pellets. This process requires heating the catalysts to 160°C in a drying oven for 2 hours. After cooling, the reactivated catalysts can be stored in a dry area until further use. Reduced conditions in the jar are monitored by the addition of methylene blue or resazurin indicators. These indicators, initially blue and pink (respectively), become colorless with low concentrations of oxygen.

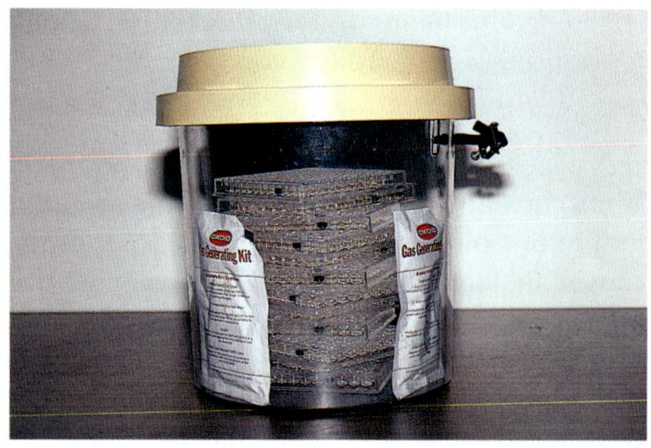

Figure 2-5. Large-size GasPak jar (Becton Dickinson Microbiology Products) for incubating MIC trays or three stacks of plates.

Anaerobic jars can be set up by two different methods. The easiest employs a commercially available gas generator envelope which releases H_2 and CO_2 when 10 ml of water is added. The envelope is placed into the jar (equipped with a catalyst and indicator), water is added, and the jar is sealed. Production of heat within a few minutes (detected by touching the lid or the outside of the jar near the catalyst) and subsequent development of moisture on the walls of the jar are indications that the catalyst and generator envelope are functioning properly. Anaerobiosis is achieved in 1 to 2 hours, although the indicators may take longer to decolorize.

Alternatively, the "evacuation-replacement" system may be used. Air is removed from the sealed jar by drawing a vacuum of 25 inches of mercury. This process is repeated two times, filling the jar with an oxygen-free gas such as N_2 between evacuations. The final fill of the jar is made with a gas mixture containing 80 to 90% N_2, 5 to 10% H_2, and 5 to 10% CO_2. Once again, an indicator is included to indicate anaerobiosis. Anaerobiosis is achieved more quickly by the "evacuation-replacement" method. Both methods give comparable yields of anaerobes from clinical specimens, if the specimen is properly transported and set up in jars immediately after plates are streaked.

The Anoxomat (Mart BV[R] Microbiology Automation) is another method that employs the evacuation-replacement technique to create an anaerobic environment in the jar.[29] This fully automated system is similar to the technique described before but it also provides an internal quality assurance program that executes additional tests when selected. These tests include: leak checks, a test for sufficient catalyst activity, and a test for adequate addition of replacement gas. Preliminary studies in our laboratory comparing the incubation of primary plates in the Anoxomat jars, the conventional anaerobe jars, and the anaerobic chamber have shown a slight increase in the recovery of anaerobes with the Anoxomat system compared to the other two systems.[225] This impression remains to be confirmed.

Some laboratories may modify the jar technique to create a type of "holding jar" for multiple primary plates. Essentially, the holding jar is equipped with a vented lid. After the plates have been added to the jar, the jar is filled and maintained with a continuous slow stream of oxygen-free gas. Nitrogen gas is preferred over CO_2 as the latter lowers the pH of the agar. A holding jar is left at room temperature until it is completely filled. We are skeptical about survival of delicate anaerobes in such a jar. If you must utilize this type of system, we suggest that you use an oxygen-sensitive organism such as *Fusobacterium nucleatum* to evaluate anaerobiosis and that you close and incubate the jar within two hours to facilitate anaerobic recovery.

Anaerobic bag techniques.   There are two anaerobic bag systems commercially available in the U.S.: the GasPak Pouch (BBL Microbiology Systems) and the Bio-Bag Environmental Chamber Type A (Becton-Dickinson Microbiology Systems). The GasPak Pouch consists of a transparent bag (bar or heat sealed), a catalyst-free liquid reagent packet (anaerobic atmosphere), and a methylene blue indicator strip. The Bio-Bag consists of a transparent bag (heat sealed), an anaerobic gas generator ampule, a catalyst, and a resazurin indicator (Figure 2-6). An anaerobic bag system that utilizes wetted (rusting) iron filings to remove oxygen from the bag atmosphere is available in Europe (Merck/Darmstadt).

It has been shown that if an anaerobic chamber is not available for use, an anaerobic bag can be used to incubate single plates (primary and subculture) with good anaerobic recovery.[61] This study showed that the Merck-type anaerobic bag was even more effective than the Bio-Bag.

Roll-tube technique.   The roll-tube method, developed originally by Hungate [113] and popularized for clinical work by Holdeman and Moore at the Virginia Polytechnic Institute,[109] is useful for recovery of strict anaerobes when an anaerobic chamber is not available. All preparations and manipulations of the pre-reduced, anaerobically sterilized (PRAS) media-containing tubes are carried out under a constant flow of oxygen-free gas, usually introduced through the mouth of the tube via a bent metal cannula or blunt-ended needle. After autoclaving, the tubes of molten agar are spun rapidly in a special automated holder. Centrifugal force causes the agar to solidify evenly in a thin layer around the sides of the tube. Specimen inoculation and colony picking are performed with a bent wire, inserted through the mouth of the tube while oxygen-free gas flows into the tube. The need for specialized equipment, the high level of patience and skill required by users, and the inability to recognize subtle differences in colonial morphology through the agar have discouraged most microbiologists from employing this technique. In addition, most clinically significant anaerobes are not extremely oxygen-sensitive and can be recovered with less stringent methods.

Anaerobic chamber techniques.   Either flexible plastic bags [5] or rigid gas-tight cabinets can be used as anaerobic chambers. In the first system, access to the inside of the chamber is achieved by means of gloves sealed to the chamber (Coy Laboratory Products, Inc.; Forma Scientific, Inc.). The gloved chamber allows the microbiologist the convenience of "entering" and "exiting" the chamber without atmospheric replacement; however, loss of dexterity may be experienced as the gloves tend to be cumbersome. The glove-free chamber (Sheldon Manufacturing, Inc.) utilizes sleeves with cuffs that seal around the forearm and requires evacuation of air in the sleeve prior to entering the chamber (by means of a foot pedal)(Figure 2-7). This leaves the hands unencumbered by gloves. The Sheldon chamber may also be transformed into a gloved chamber.

In both systems material is passed in and out of the chamber through an interchange but the glove-free system has the advantage of permitting passage of small materials through the cuff (sleeve). The interchange is a rigid compartment with an inner and outer door and is attached to the chamber by a gas-tight seal. To minimize expense the interchange may initially be evacuated and refilled by a series of flushes with an inexpensive oxygen-free gas such as N_2. The final fill is made with a gas mixture containing 5 to 10% H_2, 5 to 10% CO_2 and 80 to 90% N_2. The number of evacuations required is dependent upon the porosity of the material being processed. Three evacuations and replacements are usually sufficient.

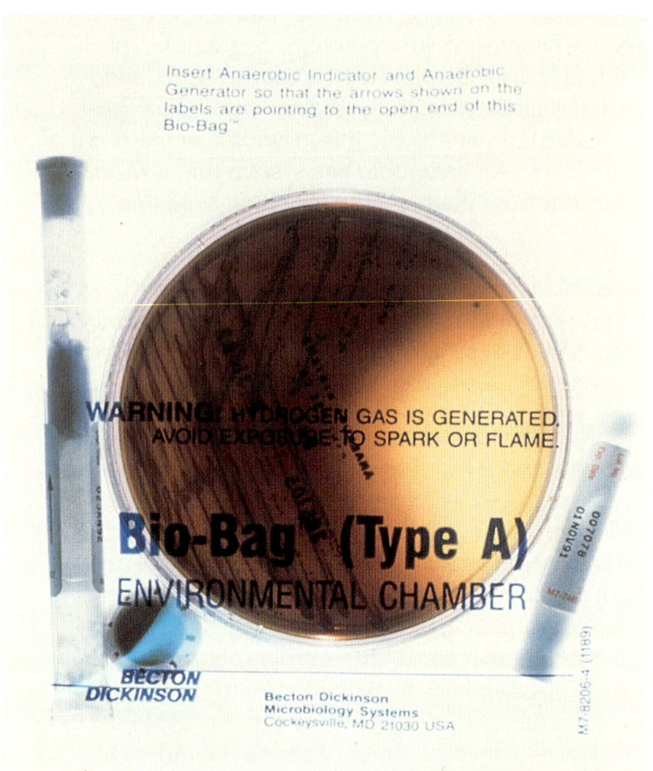

Figure 2-6. Anaerobic Bio-Bag (Becton Dickinson Microbiology Systems) containing a BBE plate growing *B. fragilis.*

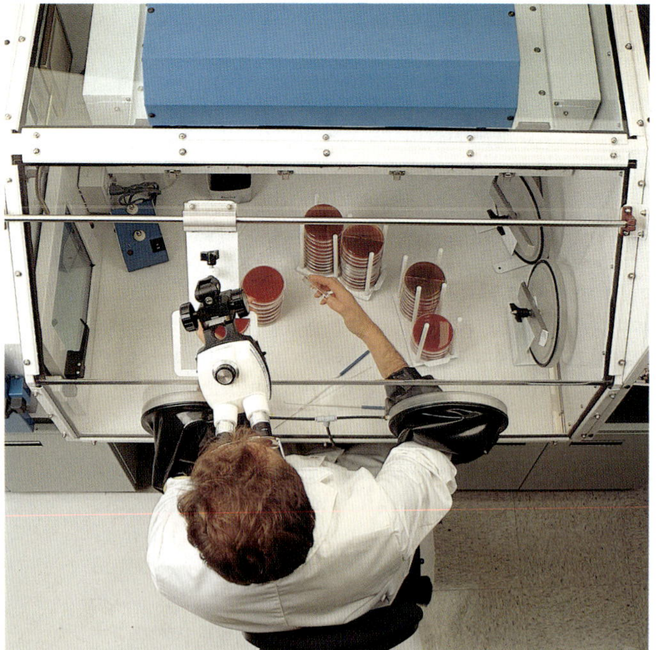

Figure 2-7. Microbiologist working in a gloveless anaerobic chamber.

The anaerobic environment of the chamber is composed of 5 to 10% H_2, 5 to 10% CO_2 and 80 to 90% N_2. Anaerobiosis is maintained by the palladium catalysts and the H_2. The H_2 is needed to rid the chamber of any oxygen that may slowly leak through the chamber walls and it also can serve as an electron donor stimulating the growth of certain anaerobes such as the *B. ureolyticus* group.[236] The CO_2 is included since many anaerobes require it for growth. The balance of N_2 is chosen for its low cost and low flammability/explosion properties.

A new methylene blue or resazurin indicator strip should be opened within the chamber every 1 to 2 days or one may open a tube of **PRAS** peptone yeast extract broth (**PY**) with resazurin indicator and cap it loosely with aluminum foil. In the gloved chambers moisture produced by the catalytic conversion of H_2 and O_2 to water is controlled by its absorption into silica gel crystals with indicator. The desiccant turns from bright blue to pink when water-logged and is "recharged" by drying in a hot oven. The Sheldon chamber has a "cold spot" which condenses excess humidity and allows the water formed to be removed through an external drain.

Chapter 3

Processing Clinical Specimens and Isolation Procedures

INITIAL PROCESSING PROCEDURES

Initial processing includes direct examination and inoculation of the specimen onto appropriate media as represented in Figure 3-1.

Direct examination

Direct examination involves macroscopic and microscopic examination of the specimen, and may include gas-liquid chromatography of the specimen and darkfield or DNA probe screening techniques. The direct examination provides immediate semiquantitative information about the types of organisms present and presumptive evidence about the presence of anaerobic bacteria (sometimes specific anaerobes). Since culture results may not be available for several days, this information is important in helping to decide the initial therapy.

Macroscopic examination. The macroscopic characteristics for which the specimen is inspected, are: 1) foul odor (due to the end-products of metabolism produced by anaerobic organisms); 2) brick-red fluorescence due to pigmented *Porphyromonas* sp. or *Prevotella* sp.; 3) black necrotic tissue or discharge which may be due to the pigmented gram-negative rods; 4) sulfur granules associated with *Actinomyces* sp. and *Propionibacterium propionicus* (formerly *Arachnia propionica*); 5) purulence; and 6) blood.

Microscopic examination. The microscopic examination should always include a Gram stain (Figure 3-2). Fix the direct smear in methanol for 30 seconds to preserve red and white blood cell morphology. Standard Gram stain procedures and reagents are used except that 0.5% basic fuchsin is substituted for safranin as the counterstain. Basic fuchsin enhances the staining of gram-negative anaerobes. The Gram stain reveals the types and relative numbers of microorganisms and host cells present, and serves as a quality control measure for the microbiologist. It can also suggest using special media. Failure to recover all the morphotypes seen on the Gram stain smear may indicate a problem in anaerobic techniques in specimen collection, transportation or processing; occasionally, it may be due to the inhibition of the organisms by residual antibiotics in the specimen.

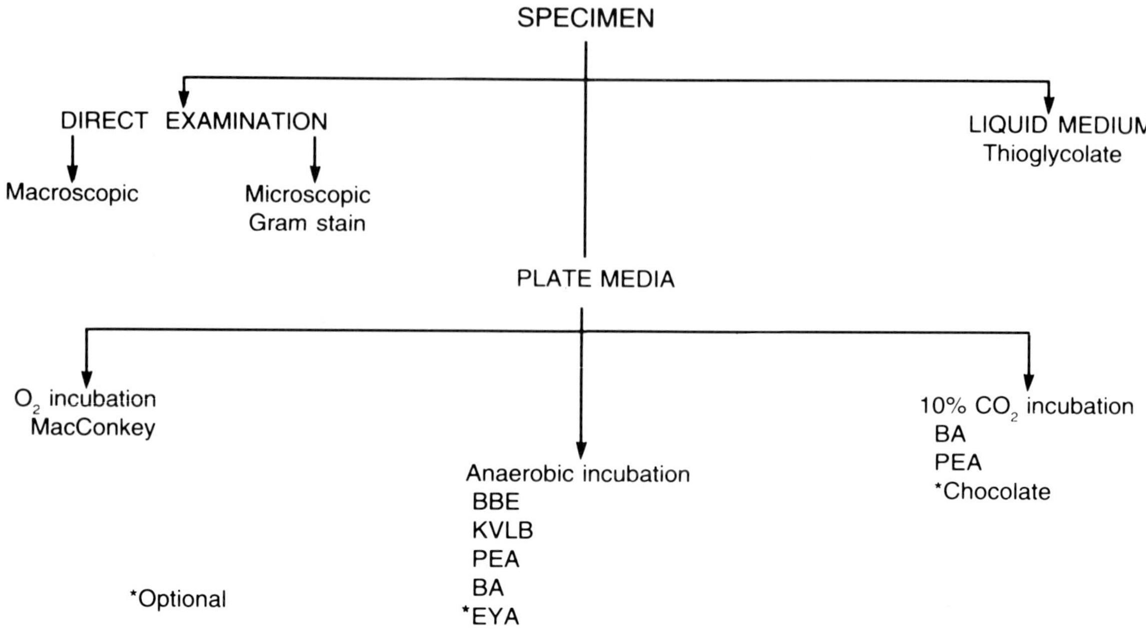

Figure 3-1.   Initial specimen processing scheme.

Pale, irregularly stained organisms are frequently anaerobic. Gram-negative, cocco-bacillary forms are suggestive for pigmented *Prevotella* or *Porphyromonas* sp. and for *Haemophilus* sp. A thin gram-negative bacillus with pointed ends suggests *Fusobacterium nucleatum,* although a new species, *F. periodonticum,* and the microaerobic *Capnocytophaga* sp. also have these characteristics. A very large gram-negative bacillus with one pointed and one square end is suggestive of *Leptotrichia buccalis* (Figure 3-3). *F. mortiferum* typically is an extremely pleomorphic gram-negative rod with filaments containing swollen areas, round bodies and irregular staining. The same type of morphology can be seen with *F. ulcerans* and occasionally with *F. necrophorum.*

A small (less than 0.5 μm) gram-negative coccus is suggestive of *Veillonella* sp. Observation of gram-positive rods with spores indicates *Clostridium* sp.; however, many clostridia appear gram-negative and spores are not always seen. A large gram-positive rod with boxcar-shaped square cells and no spores can presumptively be identified as *C. perfringens.* A branching gram-positive bacillus may be *Actinomyces, Propionibacterium, Bifidobacterium,* or *Eubacterium.* (See Figures in Chapters 4 and 5 for different cellular morphologies.)

It is always important to correlate the specimen source with microscopic findings. Darkfield or phase contrast microscopy may aid in detecting poorly stained organisms and spores, and also in observing motility directly.

Several nucleic acid probes have been manufactured and subsequently evaluated; however, they are not used widely yet.

Direct gas-liquid chromatographic analysis of clinical specimens or blood culture bottles for metabolic end-products may provide rapid presumptive evidence for anaerobic infec-

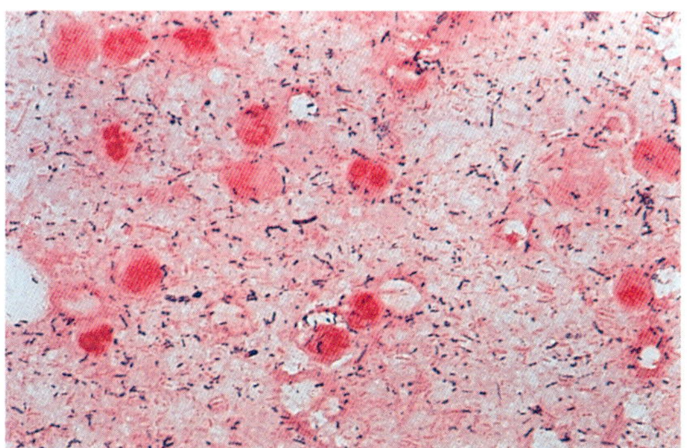

Figure 3-2. Gram stain from peritoneal fluid in a case of perforated appendicitis. Numerous bacterial morphologies are present.

Figure 3-3. Gram stain of *Leptotrichia buccalis,* showing gram-negative bacillis with one square end and one pointed end.

tion together with Gram stain and macroscopic findings.[94]   Our experience is that it seldom adds to information from Gram stain and macroscopic findings.

Specimen preparation and inoculation

After direct examination, the prepared specimen is inoculated onto appropriate anaerobic and aerobic plating media, and into liquid medium.

Specimen processing should be done in an anaerobic chamber to minimize aeration. If a chamber is not available, process the specimen within 15 minutes.

- Mix grossly purulent material by vortexing it in anaerobic conditions to ensure even distribution of the organisms.
- Homogenize pieces of tissue or bone fragments with approximately 1 ml of liquid anaerobic medium using a tissue grinder to make a thick mixture.
- Extract swab specimens in about 0.5 ml of anaerobic broth and then treat them as a liquid specimen.
- Using a Pasteur pipet, add the prepared specimen as follows:
  1) 1 drop per plate for purulent material
  2) 2 to 3 drops per plate for nonpurulent material
  3) 1 drop onto a slide for Gram stain; spread evenly
  4) the rest into liquid medium (such as thioglycolate) for enrichment culture

Media.   Use a combination of enriched, nonselective, selective and differential media for isolation and presumptive identification of bacteria from clinical material. To determine the appropriate media required for the recovery of aerobic and facultative bacteria from various sites, refer to the Manual of Clinical Microbiology, Bailey and Scott's Diagnostic Microbiology [9,10] or any other standard reference source. Sheep blood, chocolate, MacConkey and phenylethyl alcohol blood agar plates can be used as a minimum set for primary inoculation for recovery of aerobic bacteria.

The following media are used in the Wadsworth Clinical Anaerobic Bacteriology Laboratory for the isolation of the obligately anaerobic bacteria:

1) Brucella agar (BA) supplemented with 5% sheep blood and vitamin K_1 (1 μg/ml) and hemin (5 μg/ml) as a nonselective medium to support the growth of all bacteria

2) Bacteroides bile esculin (BBE) agar for the selective isolation of the *Bacteroides fragilis* group [150] and *Bilophila* [11]
3) Kanamycin-vancomycin laked blood (KVLB) agar for the selection of pigmented and other *Prevotella* sp. and *Bacteroides* sp. [79]
4) Phenylethyl alcohol sheep blood agar (PEA) to inhibit facultative gram-negative rods and to inhibit swarming of certain clostridia
5) Thioglycolate medium without indicator, supplemented with vitamin K_1, hemin, and a marble chip, for enrichment and back-up culture

The primary plating media ideally should be prereduced, unless freshly prepared media are used. Prereduced, anaerobically sterilized (PRAS) plate media are available commercially (Anaerobe Systems). Plate media used for subsequent subculturing do not need to be reduced.[219] Unless it is purchased in oxygen-free packages (Anaerobe Systems), boil or steam liquid medium for 5 minutes to drive off the dissolved oxygen; use it the same day.

In addition to Brucella agar base, brain heart infusion supplemented with 0.5% yeast extract, CDC, Columbia, FAA (Fastidious Anaerobe Agar, Lab M, England), or Schaedler agars can be used as a base for anaerobic blood agar plates. The following supplements must be added to the base medium to enhance the growth of anaerobic bacteria:[219] 1) 5% sheep, horse or rabbit blood; 2) vitamin K_1 ($1\,\mu g/ml$); and 3) hemin ($5\,\mu g/ml$). The growth of anaerobes may vary according to the basal medium used.[179] It has been reported that Brucella agar supports the growth of anaerobic gram-negative bacilli better than CDC or Schaedler and that gram-positive cocci grow in higher numbers on CDC agar than on Brucella agar.[209] The FAA medium allows fusobacteria especially to grow luxuriantly.[26,27] In our experience, brain heart infusion agar fails to support the growth of many anaerobic organisms, including some strains of *Porphyromonas* sp.

BBE agar is useful for rapid isolation and presumptive identification of the *B. fragilis* group. The medium contains $100\,\mu g/ml$ of gentamicin to inhibit most aerobic organisms, and 20% bile to inhibit most other anaerobes except for the *B. fragilis* group. As differential agents, esculin and iron have been incorporated into the agar to aid in detecting the esculin-positive *B. fragilis* group organisms. BBE agar is also a useful medium for isolation of *Bilophila* species, which usually produce black colonies on this medium due to H_2S production. Other organisms that may grow on this medium are occasional strains of *Fusobacterium mortiferum* and *F. varium*, gentamicin-resistant Enterobacteriaceae, enterococci, pseudomonas, staphylococci, and yeast. However, the colony size of the facultative anaerobic organisms is usually less than 1 mm in diameter. KVLB agar is useful for the rapid isolation of *Prevotella* sp. Of course, *B. fragilis* group strains also grow well on this medium. The medium contains $75\,\mu g/ml$ kanamycin, which inhibits most aerobic, facultative, and anaerobic gram-negative rods except for *Bacteroides* and *Prevotella,* and $7.5\,\mu g/ml$ vancomycin, which inhibits most gram-positive organisms. The laked blood allows earlier pigmentation of *Prevotella* sp. *Porphyromonas* sp. are fairly susceptible to vancomycin and may not grow on this medium. Yeast and other kanamycin-resistant organisms sometimes grow on KVLB; therefore, one should perform a Gram stain and determine the aerotolerance of all isolates.

PEA should be inoculated for purulent specimens and when mixed infections are suspected. PEA inhibits facultative gram-negative rods, preventing enterics from overgrowing the anaerobes and inhibits *Proteus* from swarming. It also prevents certain clostridia, such as *C. septicum,* from swarming, which may allow the isolation of other colonies. Most gram-positive and gram-negative anaerobes will grow on primary **PEA** medium, especially in mixed culture, and the morphology of the colonies is similar to that on blood agar plates; however, it may require a longer incubation time to detect the more slowly-growing and pigmented anaerobes.

The thioglycolate broth serves as a back-up source of culture material, which may be needed in the case of a jar failure or growth inhibition on plates due to antibiotic or other antibacterial factors (e.g., polymorphonuclear leukocytes) or small numbers of bacteria in the specimen.

If clostridia are suspected clinically or from the Gram stain of the clinical material, a primary egg yolk agar (EYA) plate can be inoculated to check for the production of lipase and lecithinase. If boxcar-shaped cells are seen in the Gram stain, a direct Nagler test can be performed (see Appendix B). If culturing for *C. difficile* (not normally recommended except for epidemiologic purposes), a CCFA (cycloserine cefoxitin fructose agar) plate should be inoculated.

For other selective and non-selective media that may be useful, see Tables A-1 to A-3 in Appendix A.

Incubation of cultures

Streak inoculated plates in a four quadrant fashion to yield isolated colonies and immediately place them in an anaerobic environment (jars, chamber, anaerobic pouches). If plates are incubated in jars, the cultures should not be exposed to air until after 48 hours incubation, since anaerobes are most sensitive to oxygen during their logarithmic phase of growth. Plates incubated in the chamber or anaerobic pouches can be inspected for growth after 24 hours or earlier, as they do not need to be removed from the anaerobic environment. Selective media, such as BBE for *Bacteroides fragilis* group or EYA for clostridia, may be examined after 24 hours of incubation since the selected organisms grow rapidly; however, one should not open an anaerobic pouch or a jar containing the nonselective blood agar plate before 48 hours.

Hold negative culture plates for a minimum of seven days before final examination. Research laboratories should hold plates for two weeks. Inspect the thioglycolate broth accompanying a set of negative plates daily; if it becomes turbid, subculture a few drops onto a blood agar plate (anaerobic incubation) and onto a chocolate or blood agar plate for CO_2 incubation.

EXAMINATION OF CULTURES AND ISOLATION OF ANAEROBES

Anaerobes are usually present in mixed culture with other anaerobes and/or aerobic or facultative bacteria. It is important to study all colony morphotypes, since facultative and anaerobic bacteria sometimes have very similar colony appearances. Clues to the presence of anaerobes in a culture include: 1) a foul odor upon opening an anaerobic jar or pouch; 2) more colony types present on anaerobically-incubated BA plate than on the CO_2 incubated plate; 3) growth on BBE or other highly selective agar; and 4) red fluorescing or black pigmented colonies on the KVLB or BA.

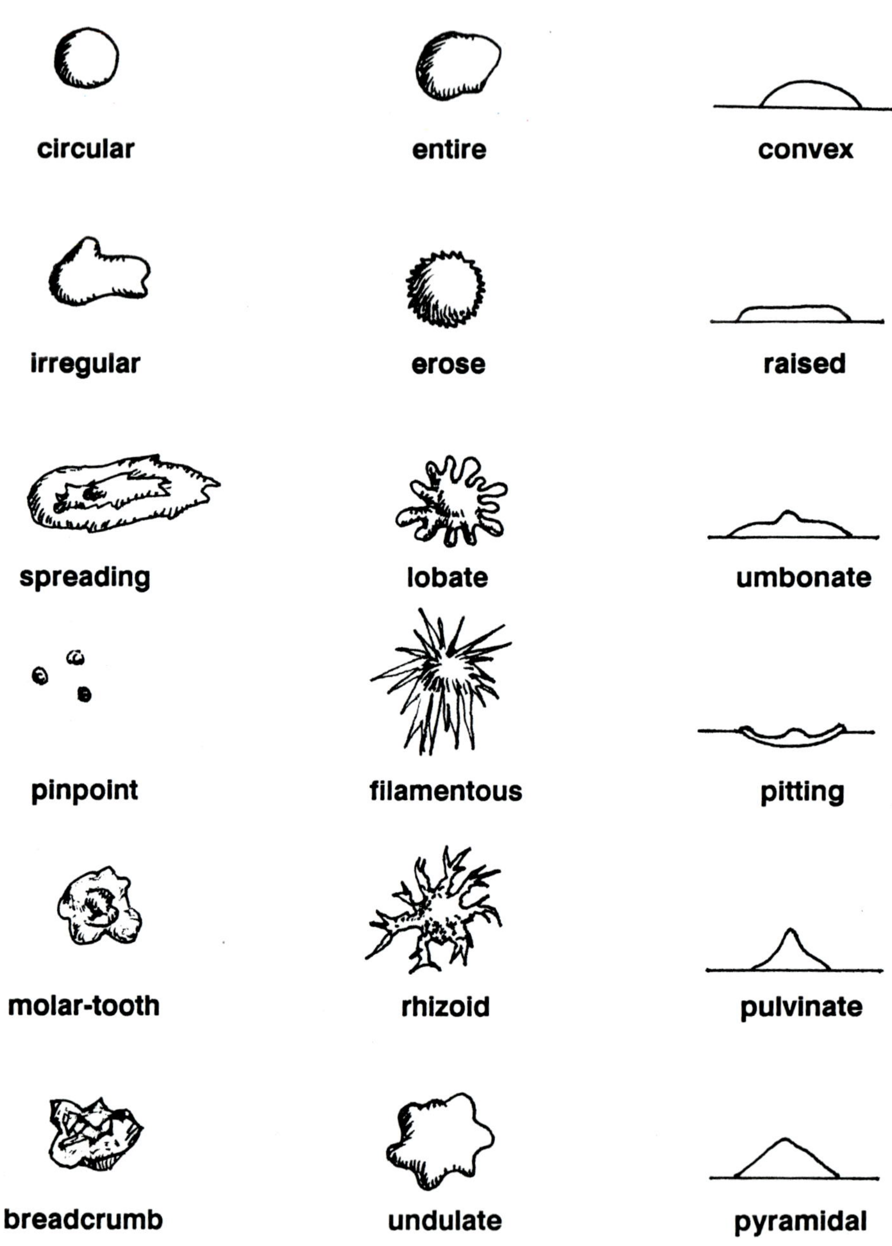

Internal characteristics include: opaque, translucent, transparent, opalescent, granular, speckled, iridescent, mucoid, glossy, matte or dull, rough, and others.

Figure 3-4.   Diagram illustrating colony morphology terminology.

Examine primary anaerobic plates with a stereoscopic microscope or a hand lens. Describe colonies from the various media, as listed below, and semiquantitate the growth. The quantity may be described as light, moderate, or heavy, or by the number of quadrants in which the growth occurs (1+, 2+, 3+, 4+ growth).

All different colony types observed should be described, semiquantitated and sub-cultured for further testing. A detailed colony description includes (Figure 3-4): 1) size; 2) shape (circular, etc.); 3) edge (entire, undulate, lobate, etc.); 4) profile (convex, flat, umbonate, etc.); 5) color; 6) opacity (transparent, translucent, opaque); 7) any other noteworthy characteristics such as pigment, fluorescence, pitting, hemolysis.

Pitting colonies on anaerobic BA plates but not on the CO_2 plate are suggestive of the *Bacteroides ureolyticus* group. Large colonies with a double-zone of beta-hemolysis are presumptively *C. perfringens.* These colonies should be Gram stained and subcultured on an EYA plate for the Nagler test. Breadcrumb-like or speckled colonies that are subsequently shown to contain slender fusiform gram-negative bacilli suggest *F. nuclea-tum.* A molar tooth-shaped colony of gram-positive branching rods may be *Actinomyces* sp. or *Propionibacterium propionicum.* The BA plate should also be examined for pigmented organisms and fluorescence under UV light (366 nm). (See Figures in Chapters 4 and 5 for different colonial morphologies on blood agar plates.)

From BBE agar, subculture the *B. fragilis* group colonies, which are typically > 1 mm in diameter, circular, entire, convex, and usually surrounded by a dark-gray zone (*B. vulgatus* is esculin-negative) (Figure 3-5). A granular precipitate around the colonies is characteristic of *B. fragilis* species. Occasional colonies of fusobacteria that grow on BBE agar are also >1 mm in diameter, but they are flat and irregular. Convex colonies on BBE with black centers and translucent margins are suggestive of *Bilophila* and should be subcultured (Figure 3-6).

Subculture all colony types from the KVLB plate (Figure 3-7) and check them carefully for pigment in bright light and for red fluorescence under UV light.

Subculture any colonies from PEA agar that have colonial morphologies not found on the BA plate; subculture all colonies from PEA if the BA plate is overgrown by swarming *Proteus,* clostridia, or other organisms.

Pleomorphic, yellow, ground-glass colonies on CCFA agar are characterictic of *C. difficile* (Figure 3-8).

Subculture of isolates

Gram stain each different colony type and subculture onto the following media (Figure 3-9):

1) BA - anaerobic incubation
2) Chocolate agar - CO_2 incubation (aerotolerance)
3) BA - O_2 incubation (optional for aerotolerance)
4) EYA for suspected *Clostridium* sp., *Fusobacterium* sp. and *Prevotella* sp. - anae-robic incubation (optional)

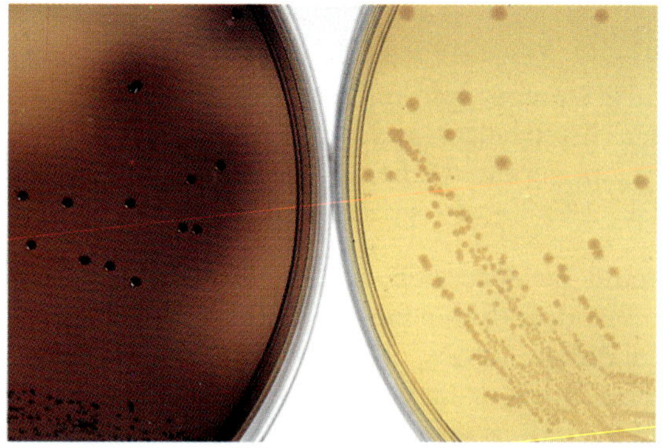

Figure 3-5. *Bacteroides fragilis* (left) vs. *B. vulgatus* on BBE agar; note the difference in esculin reaction (blackening).

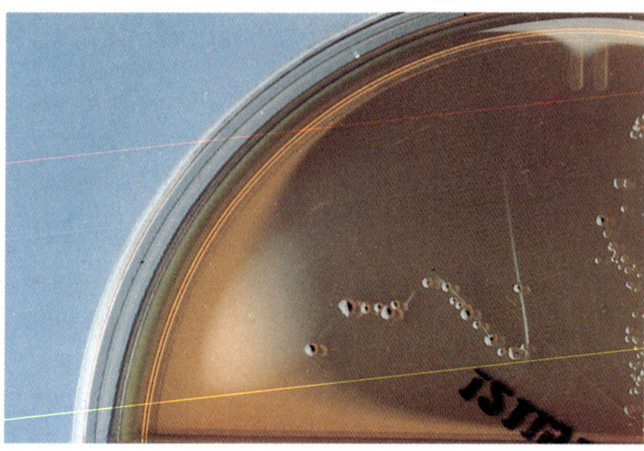

Figure 3-6. *Bilophila wadsworthia* on BBE agar; note the black-centered colonies.

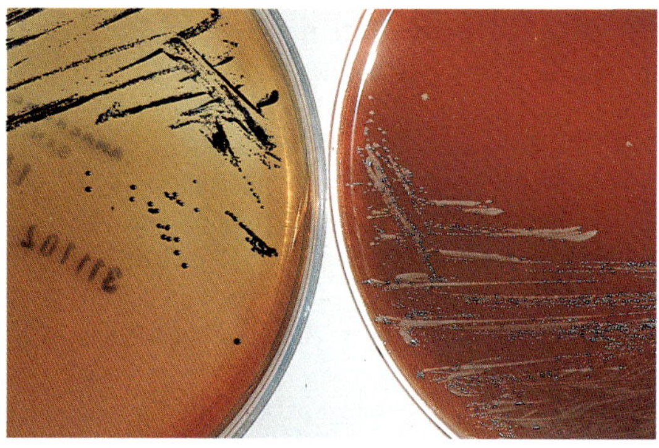

Figure 3-7. *Prevotella intermedia* on KVLB (left) and Brucella blood agar (right); note the stronger pigment production on KVLB.

Figure 3-8. *Clostridium difficile* on CCFA agar; note the yellow, ground-glass appearance of colonies and pale pink color of uninoculated medium.

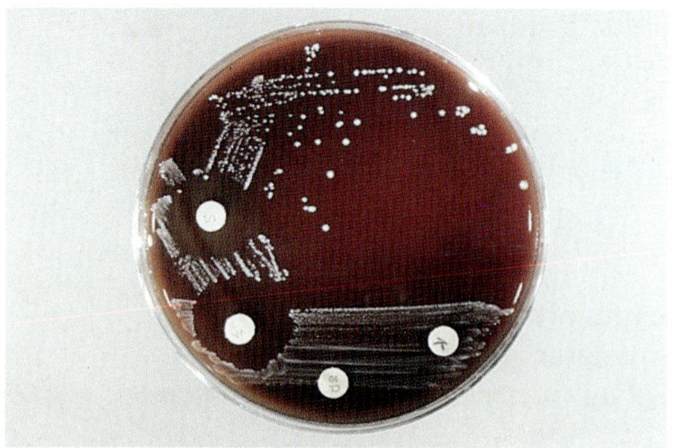

Figure 3-10. Special potency disks arranged on subculture of single colony from primary plate.

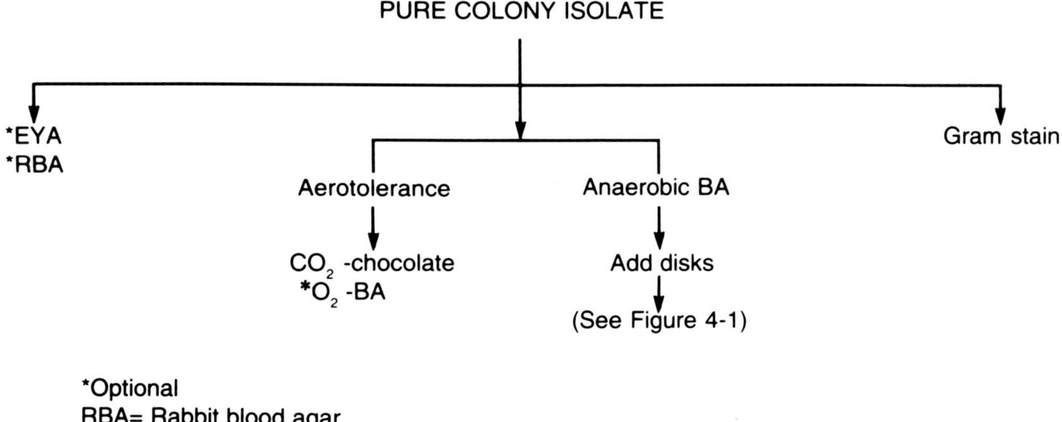

Figure 3-9.   Flow chart illustrating scheme for subculturing isolated colonies.

5)   Rabbit laked blood agar (RBA) for enhanced detection of pigmented organisms—anaerobic incubation (optional)

Try to use only one colony for all subculturing, aerotolerance testing, and Gram stain. Use a sterile wooden stick for transferring, because it picks up the colony and retains enough material for inoculation of multiple plates and for Gram stain. If the original colony is not large enough to inoculate all the plates, inoculate only the anaerobic BA plate and use that subculture for subsequent testing. EYA, RBA, and the aerotolerance plates can be divided into wedges so that several isolates can be inoculated onto each plate.

Streak the first and second quadrants of the anaerobic BA plate back and forth several times to ensure an even lawn of heavy growth; streak the other two quadrants for isolated colonies. Place the special potency disks (kanamycin, 1,000 μg; colistin, 10 μg; and vancomycin, 5 μg) on the first quadrant (Figure 3-10) well apart from each other. These disks serve as an aid in grouping the anaerobic bacteria and in determining the Gram reaction, but do not imply susceptibility of an organism for antibiotic therapy.[230]   A 10 μg sodium polyanethol sulfonate (SPS) disk may be placed near the colistin disk for rapid identification of *Peptostreptococcus anaerobius,*[258]   and a nitrate disk may be placed on the second quadrant for the detection of nitrate reduction.[257]   Alternatively, the latter two disks may be used on selected isolates at a later stage.

If processing is done on the open bench, incubate all anaerobic plates promptly (within 20 minutes), since some isolates may die after relatively short exposure to oxygen (e.g., *F. necrophorum, Porphyromonas* sp.). Reincubate the primary plates along with the subcultures for an additional 48 hours and inspect them again for new morphotypes and pigment production.

A chocolate agar plate incubated in 5% to 10% CO_2 atmosphere is required for aerotolerance testing to detect organisms that require CO_2, especially slow-growing, fastidious, facultative or microaerobic species that do not grow alone on blood agar plates (such as *Haemophilus*). Use of a BA plate alone for CO_2 incubation may yield false negative results. An additional BA plate incubated in air will further define the atmospheric requirements of facultative organisms.

Many gram-positive nonsporeforming rods, such as most *Actinomyces* sp., many *Propionibacterium* sp., *Bifidobacterium* sp., and *Lactobacillus* sp., can grow in CO_2-enriched ambient atmosphere and may be aerotolerant enough to grow poorly in air, but are considered to be anaerobic bacteria. Several species of *Clostridium,* including *C. carnis, C. haemolyticum,* and *C. tertium* can grow, but not sporulate, in air. Furthermore, some microaerobic streptococci require several subcultures before they can grow in a CO_2 atmosphere.

Chapter 4

Preliminary Identification Methods (Levels I and II)

GENERAL CONSIDERATIONS

Isolates in pure culture are further processed for identification as shown in Figure 4-1. The Wadsworth Anaerobic Bacteriology Laboratory, as a research facility, is committed to performing as complete an identification as possible. However, this philosophy is not practical for all laboratories for a variety of reasons, such as time, expense, material and personnel considerations.

In this Manual we describe three different levels at which laboratories can operate with regard to anaerobic workup. All laboratories should be capable of functioning at Level I: presumptive identification from primary plates, isolation, and maintenance of an anaerobe in pure culture so that the isolate can be sent to a reference laboratory for identification and susceptibility testing if necessary or desirable. Level II uses simple tests to further group and identify anaerobes. Level III laboratories identify isolated anaerobes as completely as possible using a variety of techniques that may include PRAS biochemicals, miniaturized biochemical systems (e.g., API, Minitek), rapid enzyme detection panels (e.g., RapID ANA, Vitek, Rapid ID 32A) and disks (Rosco), gas-liquid chromatography, MIS fatty acid methyl ester analysis, and toxin assays.

LEVEL I IDENTIFICATION

Information from the primary plates, in conjunction with the atmospheric requirements, Gram stain, and colonial morphology of a pure isolate, provides presumptive identification of anaerobic organisms. Table 4-1 summarizes the extent to which isolates can be identified using this information.

LEVEL II IDENTIFICATION

Preliminary grouping of anaerobes and even complete identification of some organisms are based on colonial and cellular morphology, Gram reaction, susceptibility to the

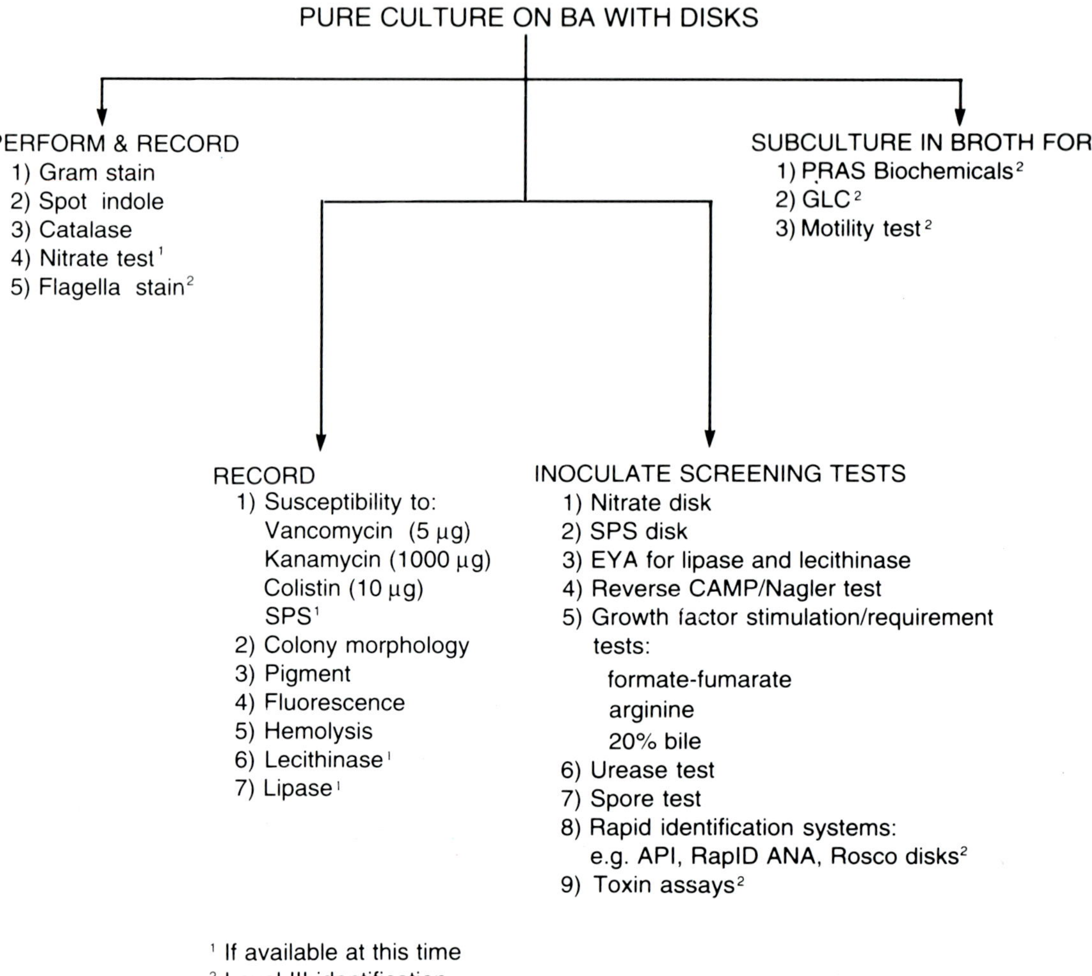

Figure 4-1. Flow chart illustrating scheme for initial identification of anaerobes from pure cultures.

special potency antibiotic disks, catalase, indole and nitrate reactions, and possibly other simple tests (Tables 4-2, 4-3). The test procedures and principles are described in Appendix B.

Preliminary examination of all isolates

The following characteristics are recorded and certain direct tests are performed from a pure culture on a BA plate for all isolates:

1) Describe colonial morphology in detail; this includes colony size, shape, edge, internal appearance, color, profile, opacity, general appearance (e.g., pitting, mucoid, dull, breadcrumb-like), and any other distinctive characteristic (Figures 3-4 to 3-8 and those in this chapter and Chapter 5). The pitting characteristic may be easier to detect after four days of incubation.

Key to tables:

Aerotolerance:

 A = anaerobic

 M=microaerobic

 F = facultative

Cell morphology:

 B = bacillus

 C = coccus

 CB = coccobacillus

Spore characteristics:

 O = oval

 R = round

 S = subterminal

 T = terminal

Susceptibility:

 R = resistant

 S = sensitive

PRAS Carbohydrates:

 pH ≤5.5 positive (+)

 pH 5.6 - 5.8 weak (W)

 pH > 5.8 negative (-)

 PYG=peptone-yeast-glucose broth

GLC = gas-liquid chromatography:

 A = acetic

 P = propionic

 IB = isobutyric

 B = butyric

 IV = isovaleric

 V = valeric

 IC = isocaproic

C = caproic

L = lactic

S = succinic

PAA= phenylacetic acid

LACTATE P = lactate converted to propionic acid

THREONINE P = threonine converted to propionic acid

 Note: capital letters indictate major metabolic

   products, small letters minor products and

   parentheses variable reaction.

Reactions:

 - = negative

 + = positive

 W = weak positive

 V = variable reaction

 +⁻ = most strains positive, some negative

 +ʷ = most strains positive, some weakly positive

 W⁻ = most strains weakly positive, some negative

 -⁺ = most strains negative, some positive

 -ʷ = most strains negative, some weakly positive

F/F REQ.= formate/fumarate required

ALK PHOS = alkaline phosphatase

Milk reactions:

c = clot formed

d = digested

d⁻ = most strains digest milk, some negative

dᶜ = most strains digest milk, some produce clot

Table 4-1. Preliminary identification

	Cell shape	Gram reaction	Aero-tolerance	Distinguishing characteristics
B. fragilis-group	B	-	-	Growth on BBE with colony size > 1mm in diameter
Pigmented *Prevotella* sp. or *Porphyromonas* sp.	CB, B	-	-	Black/brown pigmented or red fluorescing colony
B. ureolyticus group	B	-	-	Pitting colonies
F. nucleatum (presumptive)	B	-	-	Slender bacillus with pointed ends; "breadcrumb" or speckled colony
Anaerobic gram-negative bacillus	B	-	-	
Anaerobic gram-negative coccus	C	-	-	
Anaerobic gram-positive coccus	C	+	-	
C. perfringens (presumptive)	B	+	-	Double-zone of β-hemolysis; box car-shaped cells
Other *Clostridium* sp.	B	+	-	Spores seen on Gram stain
Anaerobic gram-positive bacillus	B	+	-	No box car-shaped cells; no spores

2) Describe cellular morphology including size, shape, Gram reaction, and any other distinctive characteristic (Figures in this chapter and Chapter 5).

3) Examine colonies for hemolysis on blood agar. Use transmitted light to look for β-hemolysis, a double zone of β-hemolysis (clear zone of hemolysis extending just beyond the colonies and zone of partial hemolysis extending well beyond the colony edge) (see Figure 4-2). Also look for greening of the agar due to H_2O_2 production around colonies (Figure 4-3a and 4-3b) which is apparent after exposure to air. This is most frequently seen with fusobacteria and some clostridia. Clostridial colonies are often pyramidal or rhizoid (Figures 4-4a and 4-4b).

4) Examine colonies for pigment; colors may vary from light tan to black with *Prevotella* sp. and *Porphyromonas* sp. (Figure 4-5), or pink to red with *Actinomyces odontolyticus* (Figure 4-6). To detect pigment, it is often helpful to pick up several colonies with a swab and look at the cell mass on the swab.

5) Examine colonies for fluorescence (Figure 4-7). A variety of colors may be seen such as red, orange, pink, and chartreuse (yellow-green).[25,28,215]

6) Perform catalase test.

7) Perform spot indole test (Figure 4-8).

8) Determine susceptibility to special potency antibiotic disks (vancomycin 5 μg, kanamycin 1,000 μg, and colistin 10 μg) (see Figure 3-9). The disks are used as an aid in determining the Gram reaction and in separating different anaerobic species and genera. Generally, gram-positive organisms are sensitive to vancomycin and resistant to colistin, whereas the gram-negative organisms are resistant to the vancomycin disk. The vancomycin susceptibility is especially helpful with those clostridia that consistently stain gram-negative. Rarely, however, some clostridia and lactobacilli can appear resistant to the vancomycin 5 μg disk.

Fluorescence of selected anaerobes

Organism	Color
Porphyromonas asaccharolytica-endodontalis	Red, orange
P. gingivalis	No fluorescence
Pigmented *Prevotella* sp.	Red
Non-pigmented gram-negative bacilli	No fluorescence or pink, orange, yellow
Fusobacterium sp.	Chartreuse
Veillonella sp.	Red
Eubacterium lentum	Red or no fluorescence
Clostridium difficile	Chartreuse
C. innocuum	Chartreuse
C. ramosum	Red

Special potency disk patterns

	Kanamycin 1000 μg	Vancomycin 5 μg	Colistin 10 μg	SPS 1000 μg
Gram-negative	V	R[1]	V	
Gram-positive	V	S[2]	R	
Bacteroides fragilis group	R	R	R	
B. ureolyticus group	S	R	S	
Fusobacterium sp.	S	R	S	
Porphyromonas sp.	R	S	R	
Prevotella sp.	R	R	V	
Veillonella sp.	S	R	S	
Peptostreptococcus anaerobius	R[S]	S	R	S
Other gram-positive cocci	S	S	R	R

[1] *Porphyromonas* sp. is vancomycin-sensitive.

[2] Rare strains of *Lactobacillus* sp. and *Clostridium* sp. may be vancomycin-resistant.

Table 4-2. Level II group and species identification of anaerobic gram-negative organisms

	PIGMENT	KANAMYCIN (1 mg)	VANCOMYCIN (5 µg)	COLISTIN (10 µg)	GROWTH IN 20% BILE	F/F REQUIRED	NITRATE	INDOLE	CATALASE	LIPASE	UREASE	MOTILITY
B. fragilis group	-	R	R	R	R	-	-	V	V	-	-	-
B. ureolyticus group	-	S	R	S	S	+	+	-	$-^{+}$	-	V	V
B. gracilis	-	S	R	S	V	+	+	-	-	-	-	-
B. ureolyticus	-	S	R	S	S	+	+	-	$-^{+}$	-	+	-
Wolinella/Campylobacter	-	S	R	S	S	+	+	-	-	-	-	+
Bilophila sp.	-	S	R	S	R	-	+	-	+	-	$+^{-}$	-
Fusobacterium sp.	-	S	R	S	V	-	-	V	-	V	$-^{+}$	-
F. necrophorum	-	S	R	S	S^R	-	-	+	-	+	-	-
F. nucleatum	-	S	R	S	S	-	-	+	-	-	-	-
F. mortiferum/varium	-	S	R	S	R	-	-	V	-	$-^{+}$	-	-
Porphyromonas sp.	+	R	S	R	S	-	-	+	-	-	-	-
Pigmented *Prevotella* sp.	+	R	R	V	S	-	-	V	-	V	-	-
P. intermedia	+	R	R	S	S	-	-	+	-	+	-	-
Other *Prevotella* sp.	-	R	R	V	S	-	$-^{+}$	$-^{+}$	$-^{+}$	$-^{+}$	-	-
Gram-negative cocci	-	S	R	S	S	-	V	-	V	-	-	-
Veillonella sp.	-	S	R	S	S	-	+	-	$-^{+}$	-	-	-

Table 4-3. Level II group and species identification of anaerobic gram-positive organisms

	CELLULAR MORPHOLOGY	SPORE TEST	KANAMYCIN (1 mg)	VANCOMYCIN (5µg)	COLISTIN (10µg)	SODIUM POLYANETHOL SULFONATE (SPS)	INDOLE	NITRATE	CATALASE	ARGININE STIMULATION	LECITHINASE	RED FLUORECENCE	REVERSE CAMP TEST	BOX CAR SHAPE	DOUBLE ZONE β-HEMOLYSIS	UREASE	YELLOW GROUND GLASS COLONIES ON CCFA
Anaerobic gram-positive cocci	C	-	V	S	R	V	V	-+	V								
P. anaerobius	C/CB	-	R^S	S	R	S	-	-	-								
P. asaccharolyticus	C	-	S	S	R	R	+	-	-+								
P. hydrogenalis	C	-	S	S	R	R	+	-									
Clostridium species	B	+	V	S^R	R		V	V	-		V						
Nagler-positive *Clostridium* species	B	+	S	S	R		V	V	-		+						
C. perfringens	B	+	S	S	R		-	V	-		+		+	+	+	-	
C. bifermentans	B	+	S	S	R		+	-	-		+		-	-	-	-	
C. sordellii	B	+	S	S	R		+	-	-		+		-	-	-	+-	
Nagler-negative *Clostridium* species	B	+	V	S^R	R		V	V	-		V	V					
C. difficile	B	+	S	S	R		-	-	-		-	-					+
Nonsporeforming bacilli	CB/B	-	S	S^R	R		V	V	V		V						
P. acnes	B	-	S	S	R		+-	+	+								
E. lentum	CB/B	-	S	S	R		-	+	-+	+	V						

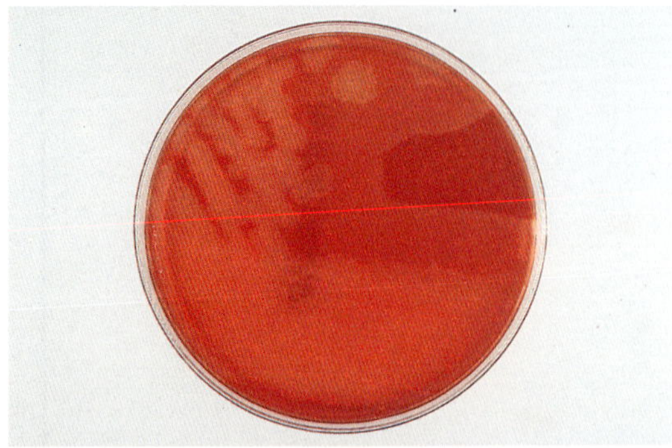

Figure 4-2. *Clostridium perfringens* on blood agar showing double zone of β-hemolysis.

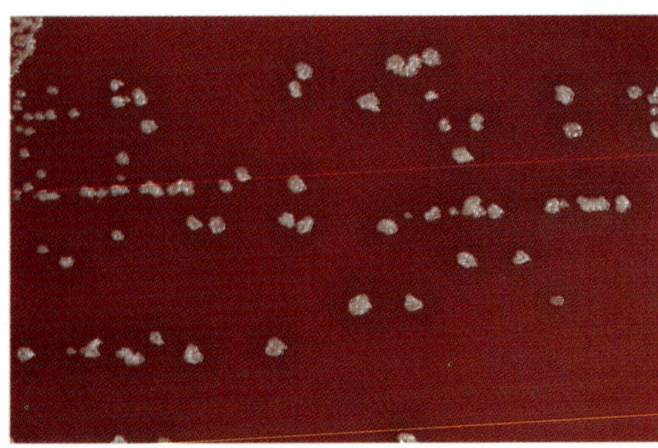

Figure 4-3a. *Fusobacterium nucleatum*: dry, irregular, white breadcrumb-like colony morphology; note greening of the agar.

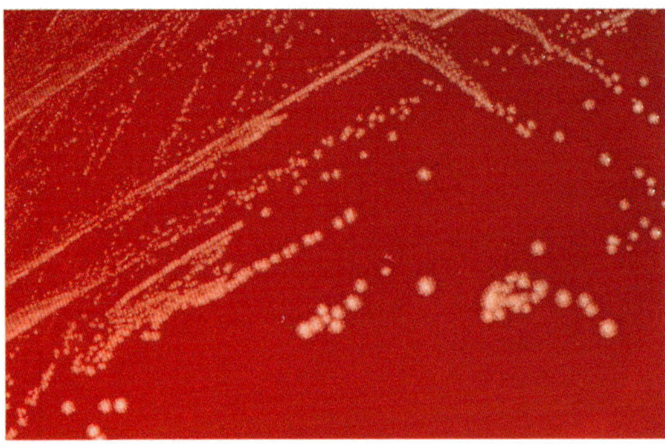

Figure 4-3b. *Fusobacterium nucleatum*: circular, entire, speckled colony morphology.

Figure 4-4a. Colony typical of clostridia on blood agar; note pyramidal shape and erose edge.

Figure 4-4b. Colony typical of clostridia on egg yolk agar; note rhizoid edge.

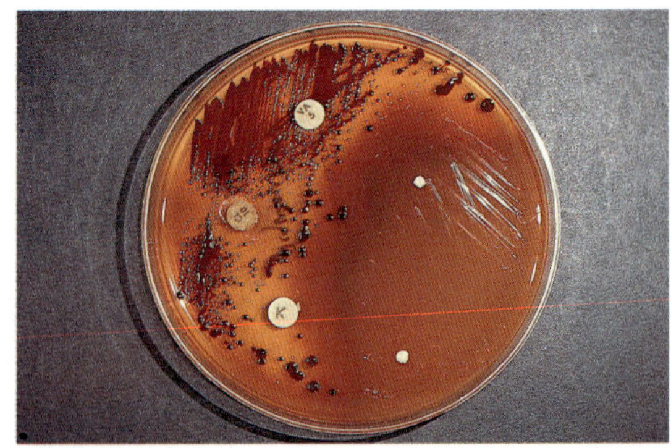

Figure 4-5. *Prevotella intermedia* on blood agar with special potency disks.

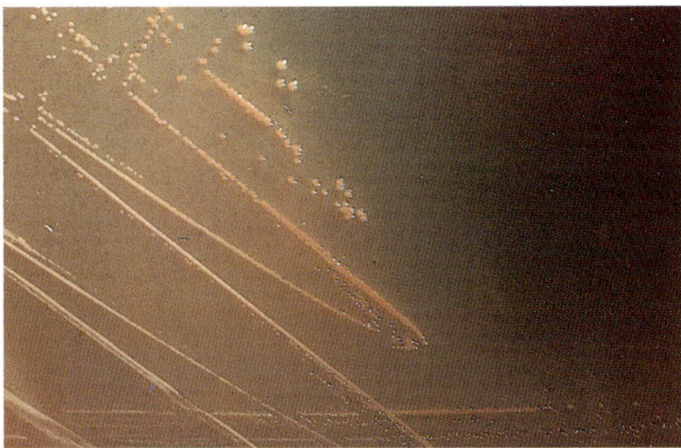

Figure 4-6. *Actinomyces odontolyticus* on blood agar; note pale orange/pink pigment.

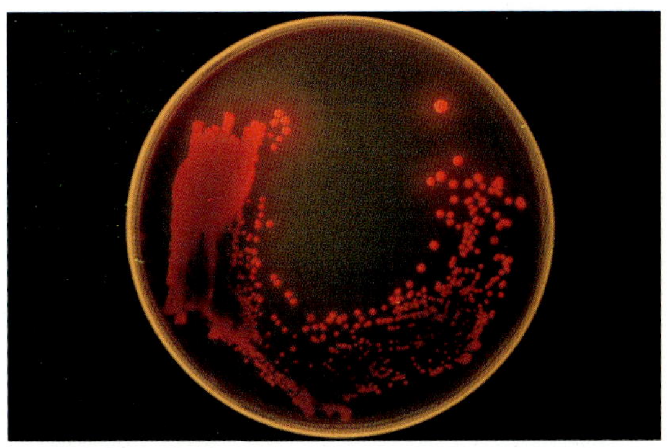

Figure 4-7. Brick-red fluorescence under long-wave UV light of pigmented anaerobic gram-negative bacillus. (Photograph courtesy of Dr. John Brazier.)

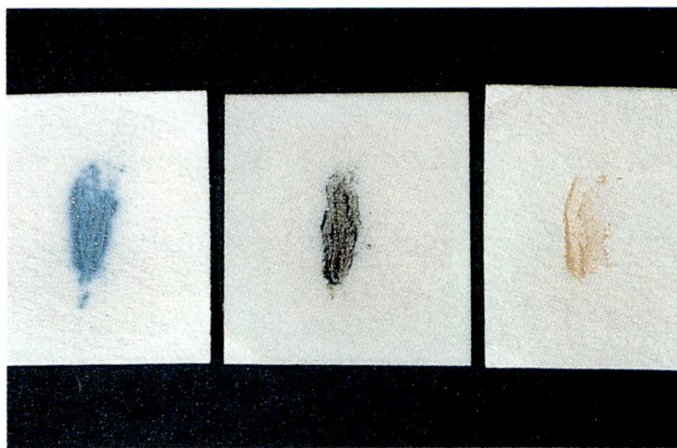

Figure 4-8. Spot indole test showing positive result on left (blue), negative result on right (pink), and slight masking of positive result by pigment of *Prevotella intermedia* (center).

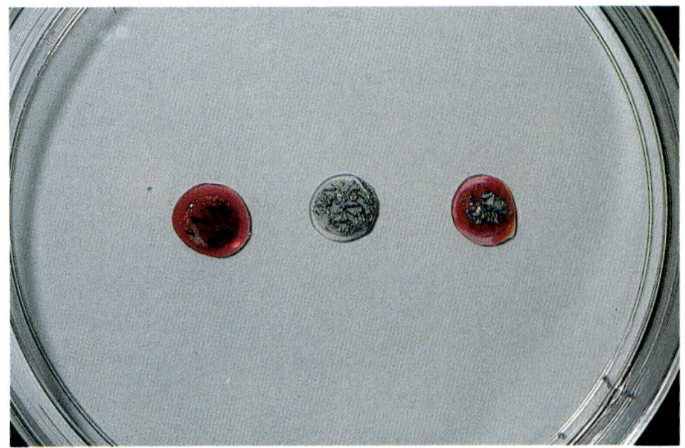

Figure 4-9. Nitrate disk test showing positive test result with reagents only (left), positive test result with reagents plus zinc (center), and negative test result with reagents plus zinc (right).

Figure 4-10. Lecithinase reaction on egg yolk agar; note opacity of agar surrounding colonies due to precipitation of complex fats.

Figure 4-11. Lipase reaction on egg yolk agar; note mother-of-pearl sheen on agar surface.

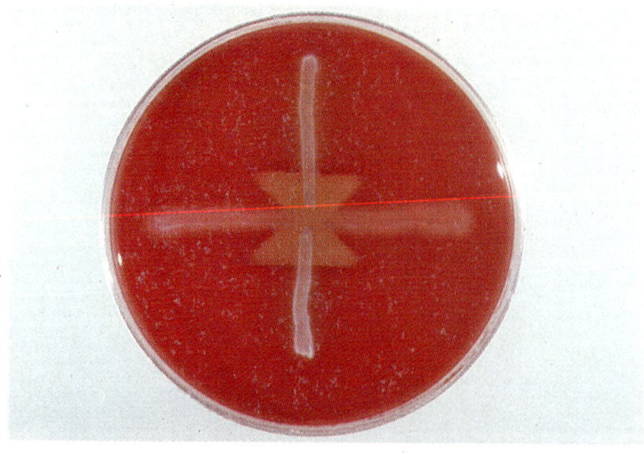

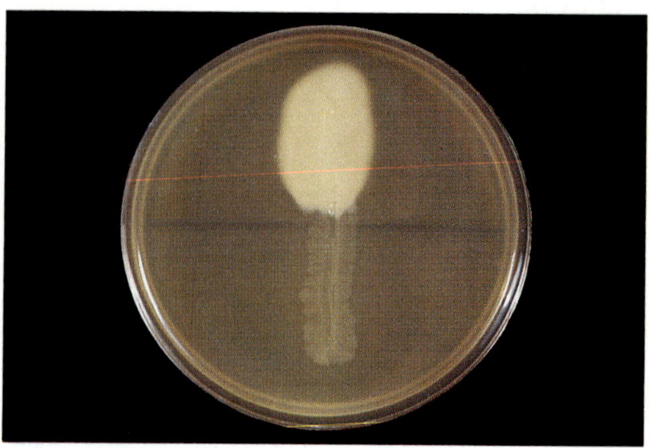

Figure 4-12. *Clostridium perfringens* (cross-streak) and *Streptococcus agalactiae* showing positive reverse CAMP test (arrow-shaped synergistic enhancement of hemolysis).

Figure 4-13. *Clostridium perfringens*: positive Nagler test.

Screening tests for selected isolates

1) Perform nitrate disk test (Figure 4-9). Test (a) indole-negative gram-positive rods (not clostridia), (b) gram-negative rods that form small transparent or pitting colonies and are susceptible to kanamycin and colistin (*B. ureolyticus*-like organisms), and (c) gram-negative cocci.

2) Determine susceptibility to sodium polyanethol sulfonate (SPS) using the disk test. Test gram-positive cocci.

3) Determine ability to produce lecithinase (Figure 4-10). Test all clostridia.

4) Determine ability to produce lipase (Figure 4-11). Test pigmented gram-negative rods, suspected fusobacteria (sensitive to kanamycin and colistin, umbonate or convex, speckled colony, foul odor), and clostridia.

5) Perform reverse CAMP test [38,102] (Figure 4-12). Test all suspected *C. perfringens*.

6) Perform Nagler test (Figure 4-13). Test all suspected *C. perfringens* (alternative to reverse CAMP test). There are four Nagler-positive clostridia including *C. perfringens* (See Table 4-3).

7) Determine ability to produce urease. Test Nagler-positive clostridia, *B. ureolyticus*-like organisms, suspected *Bilophila*, *Actinomyces* sp., and *Peptostreptococcus* sp.

8) Determine whether the organism produces spores. Test gram-positive rods and occasional gram-negative rods when needed to detect *Clostridium* sp.

9) Determine resistance to 20% bile. This test is useful for separating gram-negative bacilli: *B. fragilis* group, *Bilophila* and some fusobacteria are not inhibited by 20% bile, while *Prevotella* sp. is. If colonies were isolated from BBE agar, this test is not necessary as BBE contains 20% bile.

10) Perform formate and fumarate (F/F) growth requirement test. Test *B. ureolyticus*-like organisms.

11) Perform arginine growth stimulation test. Test small indole-negative gram-positive bacilli not resembling diphtheroids or clostridia.

12) Examine for motility. Test any curved gram-negative bacillus. Test *B. ureolyticus*-like bacilli and other gram-negative bacilli that do not key out if assumed to be non-motile.

Tables 4-2 and 4-3 group anaerobes by Gram reaction and list characteristics that identify groups and certain anaerobes. It is important for **all** characteristics to be present to assign identification.

Gram-negative bacilli.   The gram-negative bacilli are divided into the following major categories based on microscopic morphology, special potency antibiotic disk patterns, and other simple tests: 1) *B. fragilis* group, 2) *Prevotella* sp., 3) *Porphyromonas* sp., 4) *B. ureolyticus* group, 5) *Fusobacterium* sp., and 6) other gram-negative bacilli.

As discussed in Chapter 1, recent taxonomic changes have transferred *B. melaninogenicus, B. oralis* and related species into a new genus *Prevotella,*[201] and the asaccharolytic pigmented *Bacteroides* into *Porphyromonas* [207] (Tables 1-4 to 1-6).   The organisms assigned other genus designations are discussed in Level III identification. Here, "anaerobic gram-negative bacilli" is applied to those organisms not belonging to the *B. fragilis* group, the *B. ureolyticus* group, *Prevotella* sp., *Porphyromonas* sp., *Bilophila,* or *Fusobacterium* sp.

The *B. fragilis* group (Figure 4-14) can be identified by the special potency antibiotic disk pattern showing resistance to all three disks and by resistance to 20% bile in a tube test, the bile disk test, or on BBE agar.[62,251]

Anaerobic gram-negative bacilli that require formate and fumarate for growth in broth culture may be identified presumptively as belonging to the *B. ureolyticus* group.   The colonies are small, translucent or transparent, and sometimes produce greening of the agar.   Three colonial morphotypes exist: smooth and convex, pitting (Figure 4-15), and spreading, and all three can occur in the same pure culture. The organisms are sensitive to the kanamycin and colistin special potency disks, are asaccharolytic, and reduce nitrate. *B. ureolyticus* is urease-positive, *Wolinella* sp. and *Campylobacter* sp. are motile, and *B. gracilis* is urease-negative and non-motile.[236]   *W. recta* and *W. curva* were recently transferred into the genus *Campylobacter,* which also has been proposed for *B. gracilis* and *B. ureolyticus.*[247]   A recent study [101] suggests that the *B. ureolyticus* group are microaerophiles instead of true anaerobic bacteria.

Organisms that fluoresce brick-red, or produce brown to black colonies on blood-containing medium are placed into pigmented *Prevotella* sp. or *Porphyromonas* sp. (Three pigmented gram-negative rods of animal origin still remaining in the genus *Bacteroides* are discussed in Level III identification.) Indole-positive and lipase-positive pigmented coccobacilli (Figure 4-16) can be identified as *P. intermedia,* and indole-negative, but lipase-positive strains are either *P. loescheii* or *P. melaninogenica.*[4,110] Lipase-negative strains must be identified using other biochemical tests.   An indole-positive pigmenter sensitive to the vancomycin special potency disk can be identified as *Porphyromonas* sp.

Bilophila sp., which phenotypically resembles the *B. ureolyticus* group and some *Fusobacterium* sp. (Figures 4-17 and 4-18), can be distinguished by its strong positive catalase reaction and resistance to 20% bile.[11]

Fusobacterium sp. are sensitive to the special potency disks kanamycin and colistin, are catalase- and nitrate-negative, and produce a characteristic rancid odor due to butyric acid production.   *F. nucleatum* is an indole-positive thin rod with pointed ends (Figure 4-19).   It fluoresces chartreuse under UV light and produces greening of the agar upon exposure to the air.   There are three characteristic colonial morphologies of *F. nucleatum:* breadcrumb (Figure 4-3a), speckled (Figure 4-3b), and smooth. The colony sizes vary from <0.5 to 2 mm, and the color from white (breadcrumb) to gray and gray-white

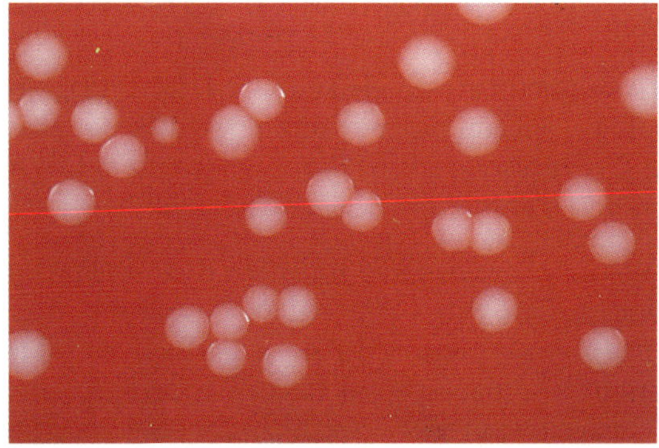

Figure 4-14. *Bacteroides fragilis* colonies on blood agar.

Figure 4-15. *Bacteroides ureolyticus* on blood agar; note pitting colony.

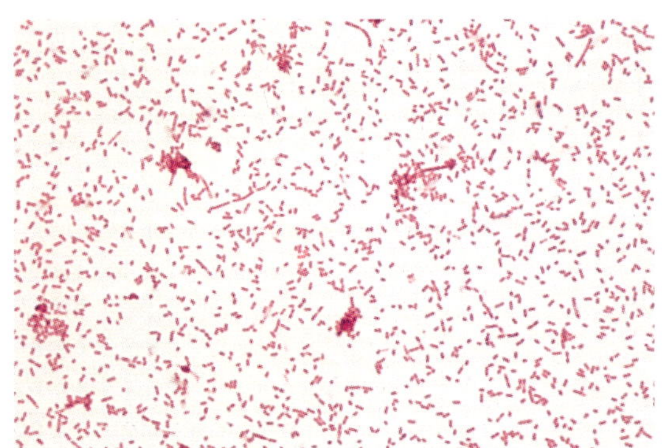

Figure 4-16. Gram stain of pigmented *Prevotella* sp.; many coccobacillary forms with some rod-shaped forms interspersed.

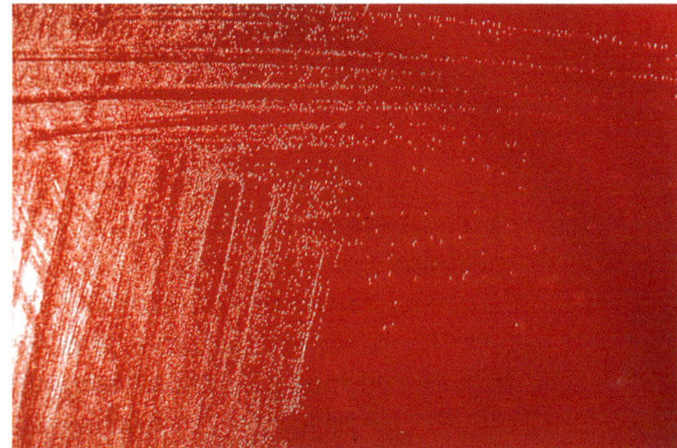

Figure 4-17. *Bilophila wadsworthia* on blood agar; colonies are small and translucent.

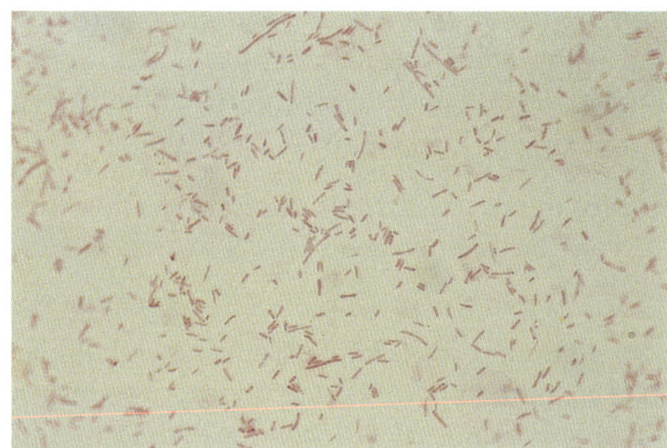

Figure 4-18. Gram stain of *B. wadsworthia* from blood agar; note straight bacillus of uniform width.

Figure 4-19. Gram stain of *Fusobacterium nucleatum*; note thin gram-negative bacillus with pointed ends.

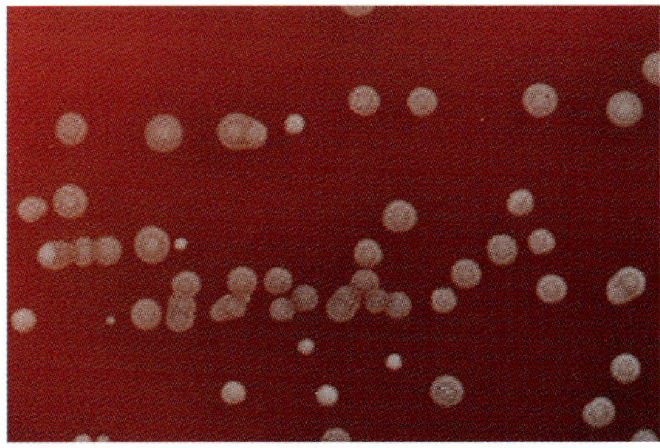

Figure 4-20. *F. necrophorum* colonies; note umbonate profile and irregular shapes.

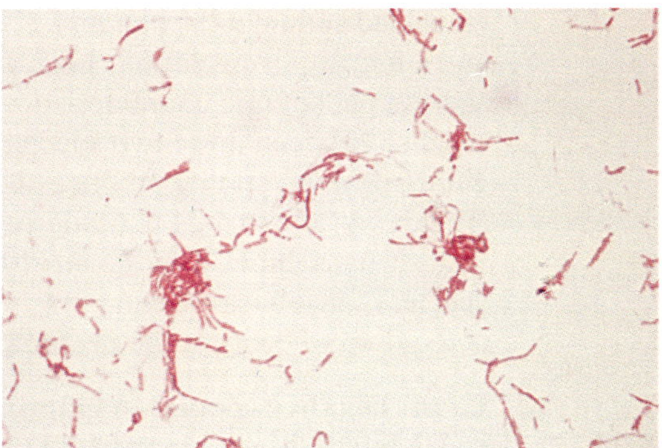

Figure 4-21. Gram stain of *F. necrophorum;* note pleomorphism and swollen forms that may be similar to *F. mortiferum.*

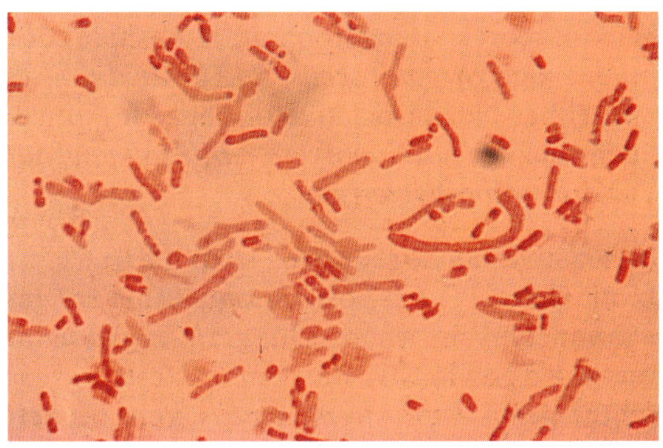

Figure 4-22. Gram stain of methanol-fixed *F. mortiferum*; note pleomorphic, swollen forms. The cell wall is so fragile that morphology may not be visualized on Gram stains prepared on heat-fixed smears.

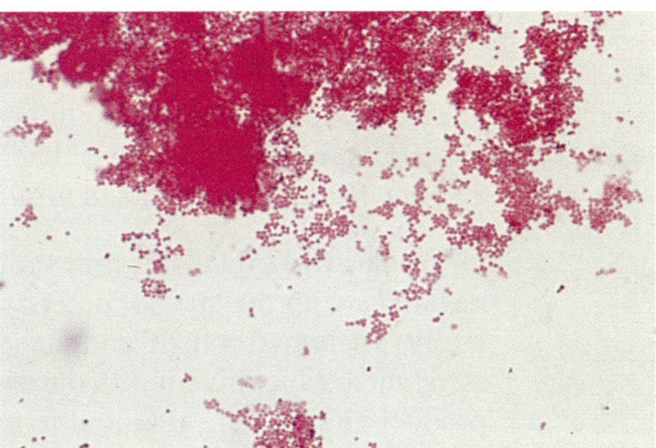

Figure 4-23. Gram stain of *Veillonella* sp.; note small gram-negative coccus.

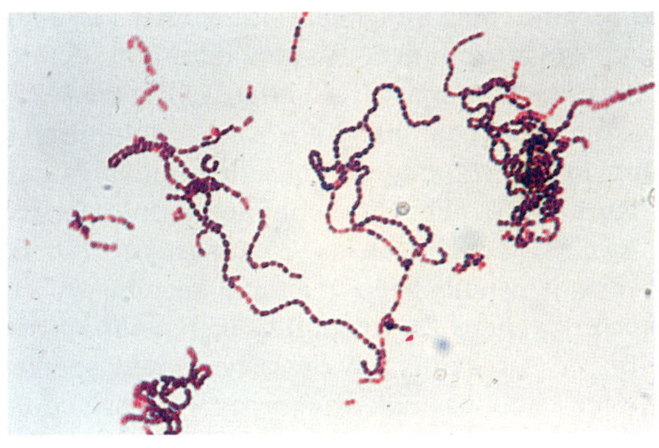

Figure 4-24. Gram stain of *Peptostreptococcus anaerobius*; large coccobacillary cells which usually form chains.

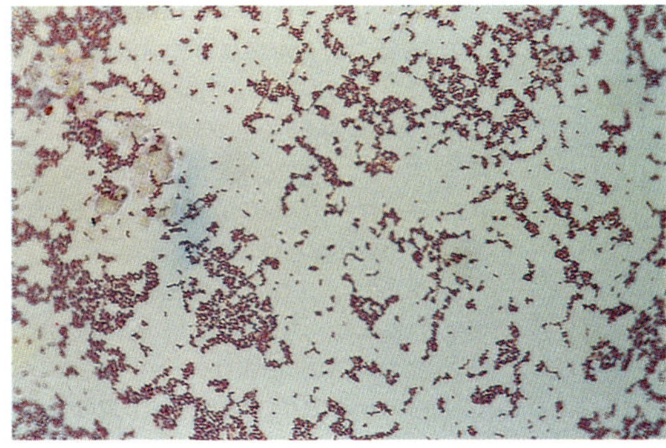

Figure 4-25. Gram stain of *P. micros*; note small size and chain formation.

(speckled and smooth colonies). *F. nucleatum* is currently divided into three subspecies,[87] and a fourth subspecies has been suggested.[88] *F. periodonticum* has been described as sharing phenotypic characteristics with *F. nucleatum;* it appears to be difficult to distinguish between these two species. A lipase-positive and indole-positive, pleomorphic gram-negative rod that produces ***b***-hemolytic colonies on blood agar is *F. necrophorum* (Figures 4-20 and 4-21). A bile-resistant *Fusobacterium* isolated from BBE agar can be tentatively identified as *F. mortiferum/varium* group (Figure 4-22), but since other species of fusobacteria can grow in the presence of 20% bile, further biochemical testing is required to confirm the presumptive identification.

Gram-negative cocci. A gram-negative coccus that reduces nitrate is *Veillonella* sp.; cells are <0.5 μm (Figure 4-23) and the colonies are small and transparent. All other anaerobic gram-negative cocci should be called "anaerobic gram-negative cocci."

Gram-positive organisms

The gram-positive organisms are divided into the following three major categories based on microscopic morphology and the presence of spores: 1) anaerobic gram-positive cocci, which includes *Peptostreptococcus, Gemella,* and *Streptococcus,* 2) *Clostridium,* the anaerobic sporeforming bacilli (which are either Nagler-positive or negative), and 3) anaerobic nonsporeforming bacilli which include *Actinomyces, Bifidobacterium, Eubacterium,* anaerobic *Lactobacillus,* and *Propionibacterium.*

Gram-positive organisms that stain gram-negative can be separated from the true gram-negatives with the special potency antibiotic disks. Gram-positive organisms are generally resistant to colistin and susceptible to vancomycin, whereas the gram-negatives are resistant to vancomycin with the exceptions noted previously. Biochemical tests and end-product analysis are required to separate the genus *Peptostreptococcus* from *Gemella* and *Streptococcus.* If spores are not observed in a Gram stain of the organism, the ethanol or heat-spore test will help separate clostridia from the anaerobic nonsporeforming bacilli. With the exception of *P. acnes* and *E. lentum,* end-product analysis is required for identification of the nonsporeformers.

Gram-positive cocci. A gram-positive coccus sensitive (exhibiting a zone of inhibition of ≥ 12mm about the disk) to sodium polyanethol sulfonate (SPS) is *P. anaerobius.* It is a large elongated coccus that occurs in pairs and chains (Figure 4-24). The colonies on blood agar are usually larger than most anaerobic cocci, ranging in size from 0.5 to 2 mm in diameter. They are gray-white and opaque in appearance. A sweet odor is associated with this organism due to the production of isocaproic acid. *P. micros* (Figure 4-25), along with rare other anaerobic gram-positive cocci can exhibit a zone of inhibition to SPS; however, the zone is usually <12 mm in size. *P. magnus* cells are noticably larger than those of other peptostreptococci (Figure 4-26). The most common indole-positive gram-positive coccus isolated from clinical specimens is *P. asaccharolyticus.* *P. indolicus,* another indole-positive coccus, is usually seen in animal specimens. The latter can be differentiated from *P. asaccharolyticus* by the ability of *P. indolicus* to reduce nitrate and its coagulase-positivity.[200] A recent study describes a new species, *P. hydrogenalis,* isolated from human feces and vaginal discharge.[235] This indole-positive coccus is coagulase-negative. Level III tests are needed to identify this species.

Clostridia. *C. perfringens* is a box car-shaped gram-positive rod (Figure 4-27) that is lecithinase- and Nagler-positive. Most strains show a typical double-zone of β-hemolysis which can be enhanced by exposure to cold (Figure 4-2). *C. perfringens* is usually reverse CAMP-positive (Figure 4-12), whereas other clostridia have been reported to be negative.[38,102] It is not necessary to demonstrate the presence of spores in *C. perfringens*. *C. sordelli* and *C. bifermentans* are similar in that they are both indole- and Nagler-positive. *C. sordellii* is usually urease-positive whereas *C. bifermentans* is always urease-negative and produces chalk-white colonies on EYA.

Nonsporeforming gram-positive bacilli. *P. acnes* is an indole-positive, catalase-positive, gram-positive pleomorphic coryneform rod (Figure 4-28). It is usually nitrate-positive. The colonies are initially small and white and become larger and more yellowish-tan with age. *P. acnes* may grow in a 5-10% CO_2 environment. The indole- or catalase-negative strains must be identified with more extensive biochemical tests.

E. lentum is a nitrate-positive, small gram-positive bacillus whose growth is stimulated by arginine. Some strains exhibit a red fluorescence under long-wave UV light. The mechanism for this characteristic is not known at the present time.

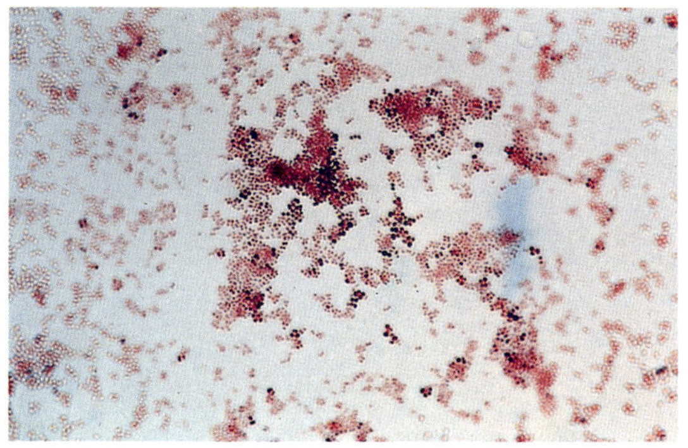

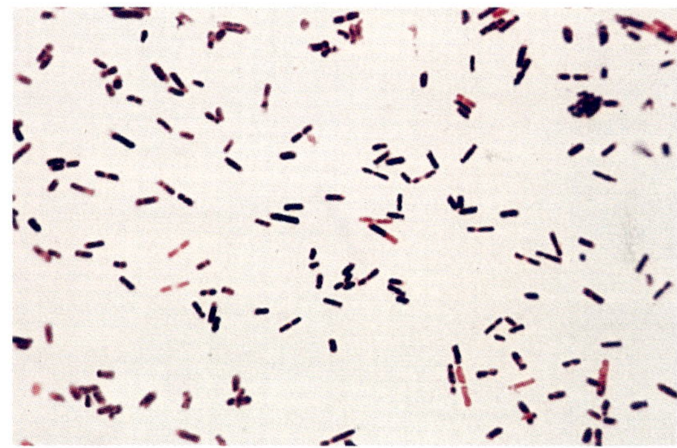

Figure 4-26. Gram stain of *P. magnus*; note large size and tetrad formation.

Figure 4-27. Gram stain of *C. perfringens*; note box car-shaped cells and absence of spores.

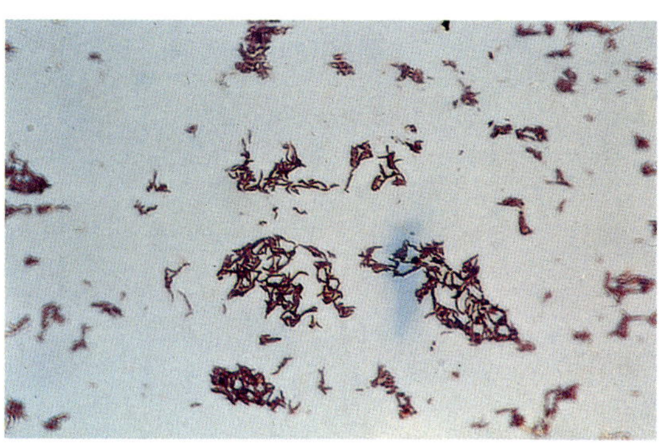

Figure 4-28. Gram stain of *Propionibacterium acnes*; note pleomorphic, branching gram-positive bacilli.

Chapter 5

Advanced Identification Methods (Level III)

GENERAL CONSIDERATIONS

Differentiation of genera of anaerobes is shown in Table 5-1. In addition to the tests performed in Level II, identification of many anaerobes to species or even genus level requires additional biochemical tests and metabolic end-product analysis by gas-liquid chromatography. In this chapter, anaerobes are divided and discussed as follows: 1) gram-negative bacilli (Tables 5-2 to 5-6 and Figures 5-1 to 5-5), 2) gram-negative cocci (Table 5-7), 3) gram-positive cocci (Table 5-8 and Figure 5-7), 4) *Clostridium* sp. (Tables 5-9 and 5-10), and 5) nonsporeforming gram-positive bacilli (Table 5-11 and Figure 5-13).

Level III identification systems are discussed at the end of this chapter and gas-liquid chromatography is covered in Chapter 7. Tables 5-2 to 5-11 and Figures 5-1 to 5-5, 5-7, and 5-13 show key characteristics for the listed organisms, based on reactions using PRAS biochemicals, gas-liquid chromatography, and preformed enzyme tests such as API ZYM and Rosco disks (see "Level III Identification Systems," below). A battery of tests to be inoculated for each anaerobe type is suggested in Table 5-12.

LEVEL III IDENTIFICATION

Gram-negative bacilli

Figure 5-1 presents a flow diagram for grouping gram-negative bacilli. *Bacteroides, Prevotella, Porphyromonas, Fusobacterium, Bilophila,* and *Campylobacter* (*Wolinella*) are encountered most commonly in clinical specimens (Table 1-2). Other gram-negative bacilli listed in Table 5-1, such as *Leptotrichia, Selenomonas* and *Anaerobiospirillum* are rare; they may occur as "contaminating" indigenous flora or occasionally as causes of infection. Bacteremia in an immunocompromised host is the most common form of infection with this latter group of organisms. Also, *Anaerobiospirillum* has been isolated from fecal specimens of patients with diarrhea,[160] and a zoonotic role for this

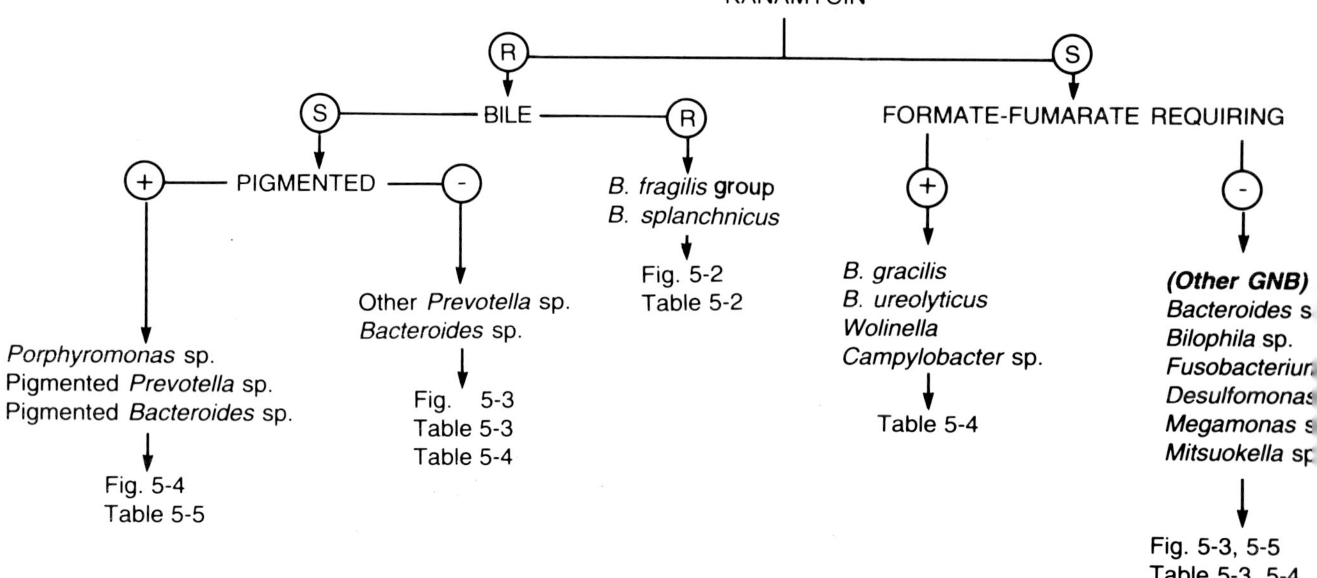

Figure 5-1.   Flow chart illustrating scheme for identification of anaerobic gram-negative bacilli.

organism has been proposed.[159]   Initial differentiation of these latter genera is based on motility, flagellar arrangement, and analysis of metabolic end-products.[111]   All gram-negative bacilli that are curved, or metabolically fastidious, or that form spreading or pitting colonies, should be tested for motility and identified according to Table 5-1. A flagella stain modified from the method of Ryu [138] is described in "Biochemical Tests and Miscellaneous Procedures" in Appendix B.

***Bacteroides fragilis* group.**   Table 5-2 (Figure 5-2) presents the key characteristics for identifying the *B. fragilis* group. Good growth on BBE medium and other 20% bile (2% oxgall)-containing media is characteristic of the *B. fragilis* group, with the exception of some *B. uniformis* strains that may grow poorly in the presence of bile. Some non-*B. fragilis* group organisms are bile-resistant; included are *B. splanchnicus, B. tectum, Mitsuokella, Bilophila,* and some fusobacteria. Not all of these, however, will grow on BBE agar. Morphologic characteristics, special potency disks, and some biochemicals will differentiate these species.

Colonies of the *B. fragilis* group on a blood agar (BA) plate are 2-3 mm in diameter, circular, entire, convex, and gray to white in color (see Figure 4-14). On BBE agar, they hydrolyze esculin, blackening the agar except for most strains of *B. vulgatus* which are esculin-negative (see Figure 3-5). The cells are uniform, pleomorphic, or vacuolized, traits that are medium- and age-dependent. Some of the species within the group are very similar biochemically and therefore difficult to differentiate. Xylan fermentation is used to separate *B. ovatus* from *B. thetaiotaomicron*.[54,196] Arabinose and cellobiose are useful in identifying *B. stercoris* and *B. uniformis;* both carbohydrates are usually fermented by *B. uniformis,* but not by most strains of *B. stercoris. B. caccae* and *B. merdae* closely resemble *B. distasonis.* Arabinose is fermented by *B. caccae,* but not usually by *B. distasonis* or *B. merdae. B. merdae* and most *B. caccae* are catalase-

Table 5-1. Differentiation of genera of anaerobes

GRAM-NEGATIVE BACILLI

I. Nonmotile or peritrichous flagella

    A. Produce butyric acid without isoacids   *Fusobacterium*

    B. Produce major lactic acid   *Leptotrichia*

    C. Produce acetic acid, reduce sulfate   *Desulfomonas*

    D. Not as above   *Anaerorhabdus*

         Bacteroides

         Bilophila

         Fibrobacter [1]

         Megamonas

         Mitsuokella

         Porphyromonas

         Prevotella

         Rikenella [1]

         Ruminobacter [1]

         Sebaldella [1]

         Tissierella

II. Motile, not peritrichous flagella

    A. Fermentative

      1. Produce butyric acid   *Butyrivibrio*

      2. Produce succinic acid

        a. Spiral-shaped cells, single polar flagellum   *Succinivibrio*

        b. Spiral-shaped cells, bipolar tufts of flagella   *Anaerobiospirillum*

        c. Ovoid cells   *Succinimonas* [1]

      3. Produce propionic and acetic acids

        a. Single polar flagellum   *Anaerovibrio* [1]

        b. Tufts of flagella on concave side   *Selenomonas*

        c. Flagella in a spiral path along cell body   *Centipeda*

      4. Produce acetic acid, twitching motility   *Mobiluncus*

    B. Nonfermentative

      1. Produce succinic acid from fumarate   *Wolinella, Campylobacter*

      2. Produce acetic acid, reduce sulfate   *Desulfovibrio*

GRAM-NEGATIVE COCCI

I. Produce propionic and acetic acids   *Veillonella*

II. Produce butyric and acetic acids   *Acidaminococcus*

III. Produce isobutyric, butyric, isovaleric, valeric and caproic acids   *Megasphaera*

Table 5-1. Differentiation of genera of anaerobes

GRAM-POSITIVE COCCI

I. Require a fermentable carbohydrate

    A. Produce butyric (plus other acids) — *Coprococcus*

    B. No butyric produced — *Ruminococcus*

II. Do not require a fermentable carbohydrate

    A. Lactic acid sole major product — *Streptococcus* / *Gemella*

    B. Not as above — *Peptostreptococcus* / *Peptococcus*

GRAM-POSITIVE SPOREFORMING BACILLI — *Clostridium*

GRAM-POSITIVE NONSPOREFORMING BACILLI

I. Produce propionic and acetic acids as — *Propionibacterium*

II. No propionic acid produced

    A. Produce acetic and lactic acid (A≥L) — *Bifidobacterium*

    B. Produce lactic acid a sole Major endproduct — *Lactobacillus*

    C. Produce moderate acetic acid plus one of the following: — *Actinomyces*

       1. Major succinic and lactic acids

       2. Major succinic acid

    D. Other: butyric ± others, acetic or no major acids — *Eubacterium*

---

[1] No known human isolates.

---

negative, whereas *B. distasonis* is usually positive. Alpha-fucosidase may be a useful reaction in differentiating *B. distasonis* and *B. merdae* from *B. caccae*: *B. distasonis* and *B. merdae* are α-fucosidase-negative and *B. caccae* is positive (data from RapID ANA chart).

As mentioned earlier, major taxonomic changes have altered the *Bacteroides* sp. other than the *B. fragilis* group. This manual will divide these organisms into pigmented and non-pigmented gram-negative rods.

Non-pigmented *Prevotella* and other gram-negative rods.

Saccharolytic.   The saccharolytic, non-pigmented gram-negative rods are listed in Table 5-3 (Figure 5-3). The key carbohydrates useful for differentiating the non-pigmented *Prevotella* sp. are arabinose, cellobiose, lactose, salicin, sucrose, and xylose; xylan is used to differentiate *P. veroralis* from similar organisms.[249] *Prevotella oris* and *P. buccae* are phenotypically and biochemically very similar, but can readily be distinguished by the α-fucosidase and β-N-acetyl-glucosaminidase tests; *P. buccae* often appears sensitive to the special potency colistin disk, while *P. oris* is resistant.[127] A rumen bacterium, *P. ruminicola*, is similar to *P. oris*. *P. zoogleoformans*, which is

Table 5-2. Characteristics of *Bacteroides fragilis* group

	GROWTH IN 20% BILE	INDOLE	CATALASE	ESCULIN HYDROLYSIS	ARABINOSE	CELLOBIOSE	RHAMNOSE	SUCROSE	TREHALOSE	XYLAN	α-FUCOSIDASE	FATTY ACIDS FROM PYG
						FERMENTATION OF						
Bacteroides caccae	+	-	-+	+	+	+⁻	+⁻	+	+	-	+	A p S (iv)
B. distasonis	+	-	+⁻	+	-+	+	V	+	+	-	-	A p S (paa ib iv l)
B. eggerthii	+	+	-	+	+	-+	+⁻	-	-	+	-	A p S (ib iv I)
B. fragilis	+	-	+	+	-	+⁻	-	+	-	-	+	A p S paa (ib iv l)
B. merdae	+	-	-+	+	-+	V	+	+	+	-	-	A p S (ib iv)
B. ovatus	+	+	+	+	+	+	+	+	+	+	+	A p S paa (ib iv l)
B. stercoris	+	+	-	+	-+	-+	+	+	-	V	V	A p S (ib iv)
B. thetaiotaomicron	+	+	+	+	+	+⁻	+	+	+	-	+	A p S paa (ib iv l)
B. uniformis	+ʷ	+	V	+	+	+	-ʷ	+	-ʷ	V	+	A p l S (ib iv)
B. vulgatus	+	-	-+	-	+	-	+	+	-	-+	+	A p S

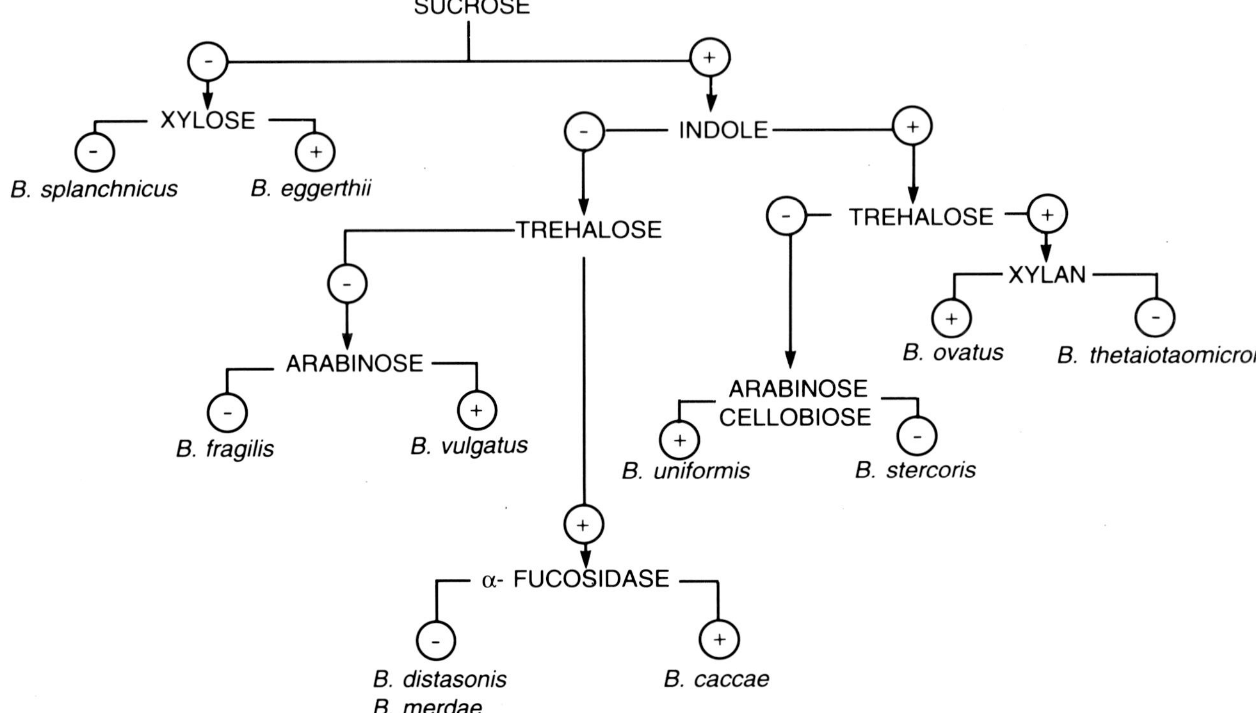

Figure 5-2.   Flow chart illustrating scheme for identification of *Bacteroides fragilis* group and *B. splanchnicus.*

infrequently isolated from clinical specimens, forms a highly viscous, tenacious (zoogleal) mass in broth cultures. Indole-positive strains of *P. zoogleoformans* were recently classified as *P. heparinolytica.*[8] *P. oulorum* is lipase-and catalase-positive. Certain strains of pigmented *Prevotella* can take up to 21 days to develop visible pigment and can therefore be difficult to differentiate from the non-pigmented species. Fluorescence is useful for recognizing these strains. *P. bivia* and *P. disiens* are strongly proteolytic and do not ferment sucrose. Lactose fermentation is used to differentiate *P. bivia* (lactose-positive) and *P. disiens* (lactose-negative). Their colonies may fluoresce pink (coral) under UV light and should not be confused with the pigmented *Prevotella* sp. *B. splanchnicus* is bile-resistant (may grow on BBE agar) and resembles the *B. fragilis* group, but does not ferment sucrose.

Asaccharolytic. Asaccharolytic, non-pigmented gram-negative rods (Table 5-4) other than the *B. ureolyticus* group organisms and *Bilophila* are occasionally isolated from clinical specimens. The recognition of *B. ureolyticus*-like organisms and *Bilophila* was discussed in Level II identification, Chapter 4. *Anaerorhabdus furcosus, Bacteroides capillosus* and *B. tectum* may ferment glucose weakly, and are esculin-positive. Gram stain and gas-liquid chromatography are useful for differentiating them (see Table 5-4). In addition, *B. capillosus* often grows better with polysorbate (Tween 80)-supplemented media and *B. tectum* is resistant to 20% bile. *B. tectum* is isolated only from animal sources or dog bite wound infections. *B. coagulans* and *B. putredinis* are both indole-positive, and can be separated by gas-liquid chromatography and the catalase test.

Table 5-3. Characteristics of saccharolytic non-pigmented gram-negative bacilli

| | GROWTH IN 20% BILE | FERMENTATION OF | | | | | | | ESCULIN HYDROLYSIS | INDOLE | α-FUCOSIDASE | β-ACETYLGLU-COSAMINIDASE | β-XYLOSIDASE | FATTY ACIDS FROM PYG |
		ARABINOSE	CELLOBIOSE	LACTOSE	SALICIN	SUCROSE	XYLOSE	XYLAN						
B. splanchnicus	+	+	+	-	+	-	-	-	+	+	+	+		A P S ib b iv (l)
Leptotrichia buccalis	+	-	+⁻	+⁻	+⁻	+	-		+	-	+	+		L (a s)
Mitsuokella dentalis	-	+	+	+	-	W	-		V	-	-	+	-	A S
M. multiacida	+	+	+	+	+	+	+		+	-				A L S
Prevotella bivia	-	-	-	+	-	-	-	-	-	-	+	+	-	A iv S (ib)
P. buccae	-	+	+	+	+	+	+	V	+	-	-	-	+	A S (p ib iv l)
P. buccalis	-	-	+	+	-	+	-	-	+	-	+	+	-	a iv s
P. disiens	-	-	-	-	-	-	-	-	-	-	-	-	-	A S (p ib iv)
P. heparinolytica[1]	-	+	+	+	+	+	+	-	+	+	+	+	+	A S (p iv)
P. oralis	-	-	+	+	+	+	-	-	+	-	+	+	-	A S (l)
P. oris	-	+⁺	+	+	+	+	+	V	+	-	+	+	+	A S (p ib iv)
P. oulorum	-	-	-	+	-	+	-	-	+	-	+	+	-	A S
P. veroralis	-	-	+	+	-	+	-	+	+	-	+	+	-	a S
P. zoogleoformans[1]	-	V	+	+	V	+	V	-	+	-		+		A P S (ib iv)

[1] Zoogleal mass produced in broth culture

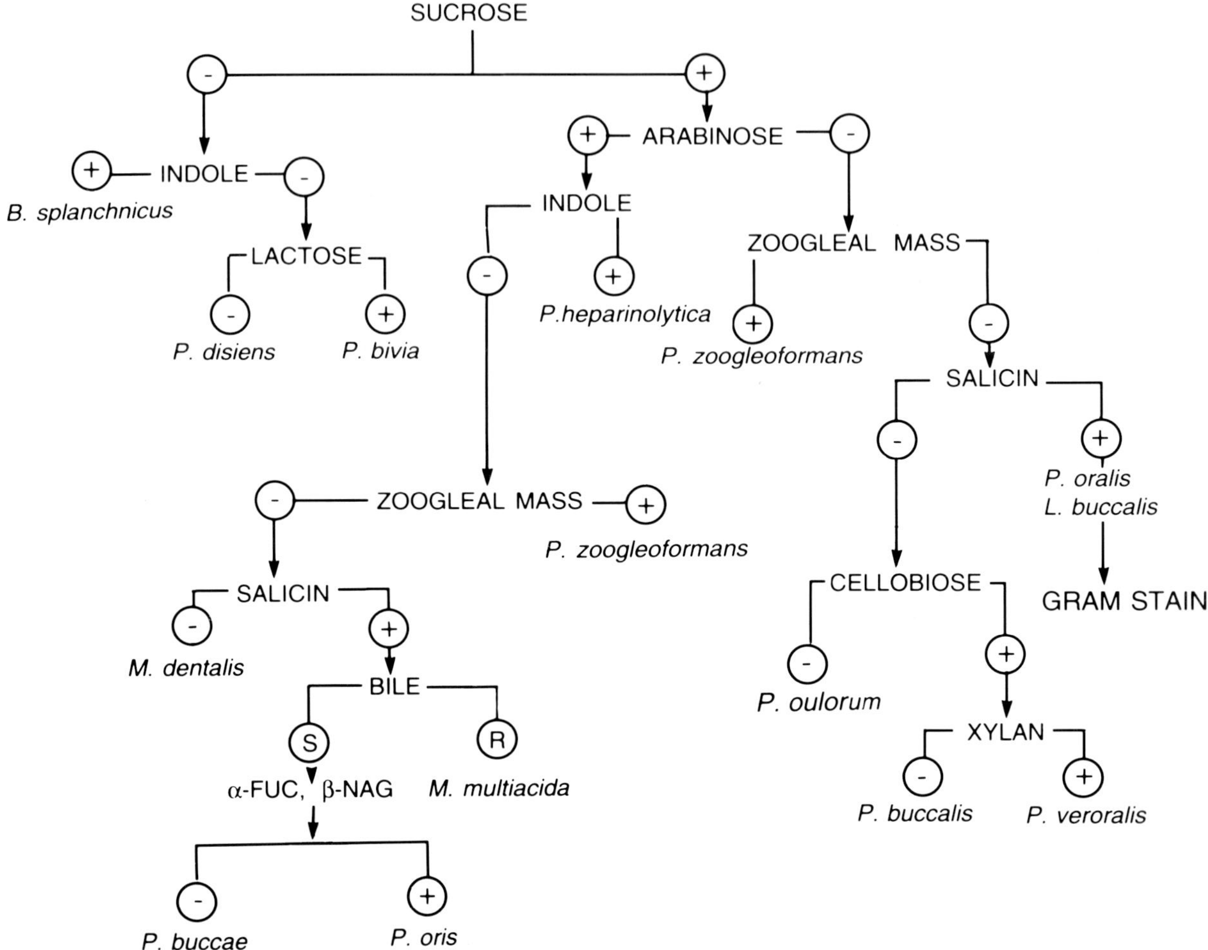

Figure 5-3. Flow chart illustrating scheme for identification of non-pigmented, saccharolytic gram-negative bacilli.

B. forsythus, which is a recognized periodontal pathogen, is a fusiform-shaped bacillus and requires N-acetylmuramic acid for growth;[267] it also shows trypsin-like activity. Visualization of *B. pneumosintes* colonies may require magnification, and negative staining may be necessary to see the cells (which are coccoid and very minute, $< 0.3 \mu$m). *Desulfomonas* is a sulfate-reducing bacterium and can be confirmed with the desulfoviridin test described in Appendix B.

Pigmented gram-negative rods. The pigmented *Prevotella* and *Porphyromonas* (and *Bacteroides*) species vary greatly in the degree and rapidity of pigmentation, depending primarily on the type of blood used. Laked rabbit blood is considered most rapid and reliable.[111] The pigmentation ranges from buff to tan to black, and may take several days to develop. The identification of those strains that usually do not show pigment until after 21 days of incubation must be established by other biochemical tests. The UV fluorescence of the pigmented *Prevotella* and *Porphyromonas* species varies from pink to orange to red, and is masked by visual pigment production. Brick-red fluorescence is the

Table 5-4. Characteristics of weakly or non-saccharolytic, non-pigmented gram-negative bacilli

	GLUCOSE FERMENTATION	INDOLE	NITRATE REDUCTION	F/F REQUIRED	UREASE	ESCULIN HYDROLYSIS	GELATIN HYDROLYSIS	CATALASE	MOTILITY	DESULFOVIRIDIN	FATTY ACIDS FROM PYG
Anaerorhabdus furcosus	W	-	-	-	-	+	-w	-	-	-	al (s)
Bilophila wadsworthia	-	-	+	-	+⁻	-	-	+	-	W⁻	A (s)
Bacteroides capillosus	W⁻	-	-	-	-	+	-w	-	-	-	a s (p l)
B. coagulans	-	+	-	-	-	-	+	-	-	-	a (p l s)
B. forsythus	-	-	-	-	-	+	+	-	-	-	A S
B. gracilis	-	-	+	+	-	-	-	-	-	-	a S
B. pneumosintes	-	-	-	-	-	-	-w	-	-	-	a (l s)
B. putredinis	-	+	-	-	-	-	+	+⁻	-	-	a P ib b IV S (l)
B. tectum[1]	W⁻	-	-	-	-	+	+	-	-	-	A P S ib iv paa
B. ureolyticus	-	-	+	+	+	-	-	-+	-	-	A S
Campylobacter/Wolinella sp.	-	-	+	+	-	-	-	-	+	-	a S
Desulfomonas sp.	-	-	V	-	V	-	-	-	-	+	A
Tissierella praeacuta	-	-	V	-	-	-	+	-	+	-	A p ib B IV s (l)

[1]Animal origin

only reliable color for presumptive identification of these organisms since some non-pigmented *Prevotella* sp. can fluoresce coral or pink. Gram stain will easily differentiate the other organisms that either produce pigment or fluoresce: the gram-positive rod, *Actinomyces odontolyticus,* usually forms red pigment; the gram-positive rod *Eubacterium lentum,* and the gram-negative coccus *Veillonella* sp. may both fluoresce red.

Table 5-5 and Figure 5-4 show characteristics for identifying the pigmented gram-negative bacilli. The unusual special potency disk pattern (sensitive to vancomycin) and asaccharolytic nature separate *Porphyromonas* from *Prevotella* sp. Phenylacetic acid production, trypsin-like activity and β-N-acetylglucosaminidase activity (see Appendix B for test procedures) will separate *P. gingivalis* from other *Porphyromonas* species; α-fucosidase activity separates *P. asaccharolytica* from *P. endotontalis.*[132,246] In addition, *P. gingivalis* does not fluoresce under UV light. *Bacteroides salivosus,* an asaccharolytic gram-negative bacillus isolated from cats and dogs,[156] phenotypically resembles *P. gingivalis*; however, the positive catalase reaction of *B. salivosus* differentiates them. *B. macacae* and *B. levii* are also of animal origin and are rarely isolated from clinical specimens. They can be separated from the human species by the positive catalase reaction of the animal strains.

Fusobacteria. *Fusobacterium* sp. are resistant to the vancomycin special potency disk and sensitive to colistin and kanamycin. They are weakly saccharolytic or non-fermentative and produce butyric acid without isobutyric or isovaleric acids. The colonial

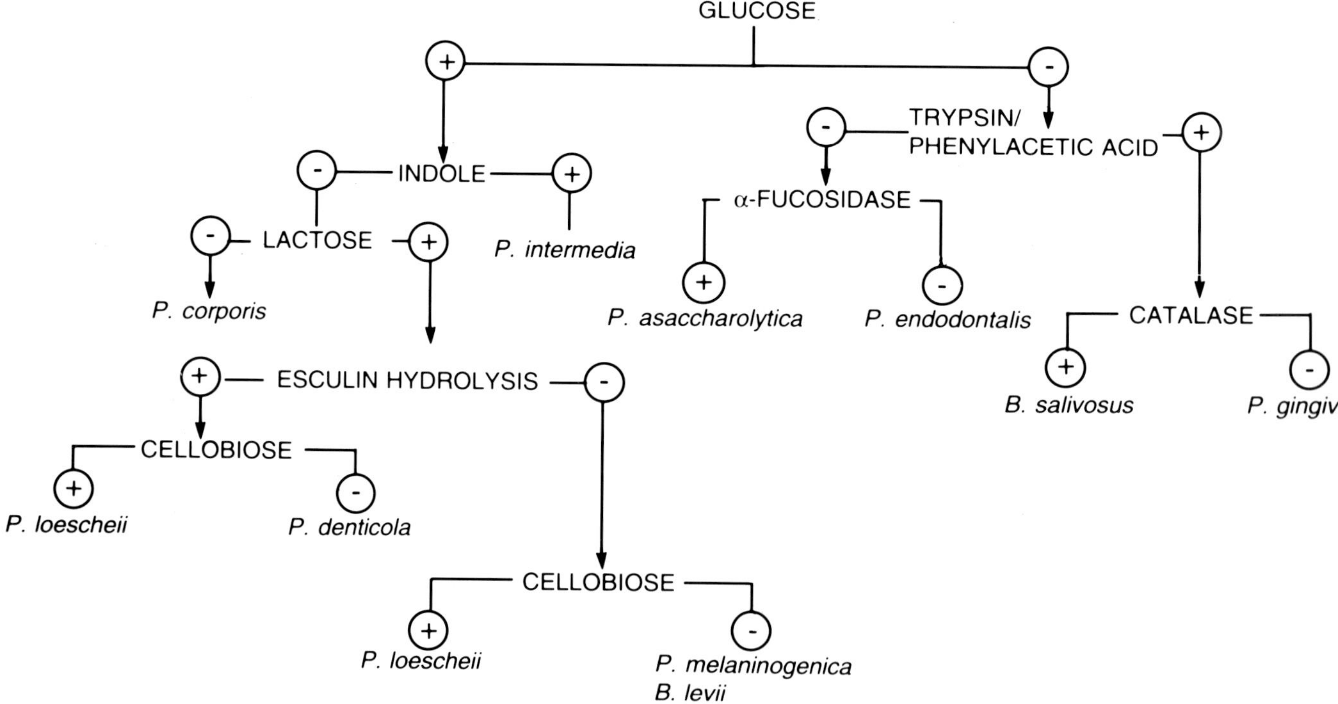

Figure 5-4. Flow chart illustrating scheme for identification of pigmented gram-negative bacilli.

Table 5-5. Characteristics of pigmented gram-negative bacilli

	INDOLE	LIPASE	CATALASE	GLUCOSE	CELLOBIOSE	LACTOSE	SALICIN	SUCROSE	ESCULIN HYDROLYSIS	α-GLUCOSIDASE	α-FUCOSIDASE	β-N-ACETYLGLUCOSAMINIDASE	TRYPSIN	FATTY ACIDS FROM PYG
					FERMENTATION OF									
Bacteroides levii[1]	-	-	V	+w	-	+w	-	-	-	V			-	A P B iv ib s
B. macacae[1]	+	-	+	+	-	-	-	-	-					a P b iv ib S
B. salivosus[1]	+	-	+	-	-	-	-	-					+	A B ib iv PAA
Porphyromonas														
asaccharolytica	+	-	-	-	-	-	-	-	-	-	+	-	-	A p ib B IV
P. endodontalis	+	-	-	-	-	-	-	-	-	-	-	-	-	A p ib B IV
P. gingivalis	+	-	-	-	-	-	-	-	-	-	-	+	+	A p ib B IV PAA
Prevotella corporis	-	-	-	+	-	-	-	-	-	+	-	+		A ib iv S (b)
P. denticola	-	-	-	+	-	+	-	+	+	+	+	+		A S (ib iv l)
P. intermedia	+	+⁻	-	+	-	-	-	+	-	+	+⁻	-		A iv S (p ib)
P. loescheii	-	V	-	+	+	+	-	+	V	+	+	+		A S (l)
P. melaninogenica	-	-+	-	+	-	+	-	+	-	+	+⁻	+		A S (ib iv l)

[1] Animal origin; catalase-negative strains with phenotypic characteristics similar to *B. levii* have been isolated from human clinical specimens. *B. salivosus* was recently moved to *Porphyromonas*.

morphology of fusobacteria varies greatly; the characteristic ones are described under Level II identification. Furthermore, many strains fluoresce chartreuse under UV light and most strains produce greening of blood agar when exposed to air due to H_2O_2 production (see Figure 4-3).

Table 5-6 lists the fusobacteria and the key reactions for differentiating them (Figure 5-5). The most common clinical isolates, *F. necrophorum* and *F. nucleatum,* are discussed under Level II identification. Lipase-negative and non-hemolytic strains of *F. necrophorum* were transferred into a new species *F. pseudonecrophorum.*[211] Bizarre, pleomorphic rods with large round bodies are characteristic of *F. mortiferum* (Figure 4-22). This organism may grow on BBE and turn the agar black. Another fusobacterium that can be isolated from BBE agar is *F. varium. F. varium* is esculin-negative and can be indole-and lipase-positive. Lipase production can take up to 10 days to develop (data not

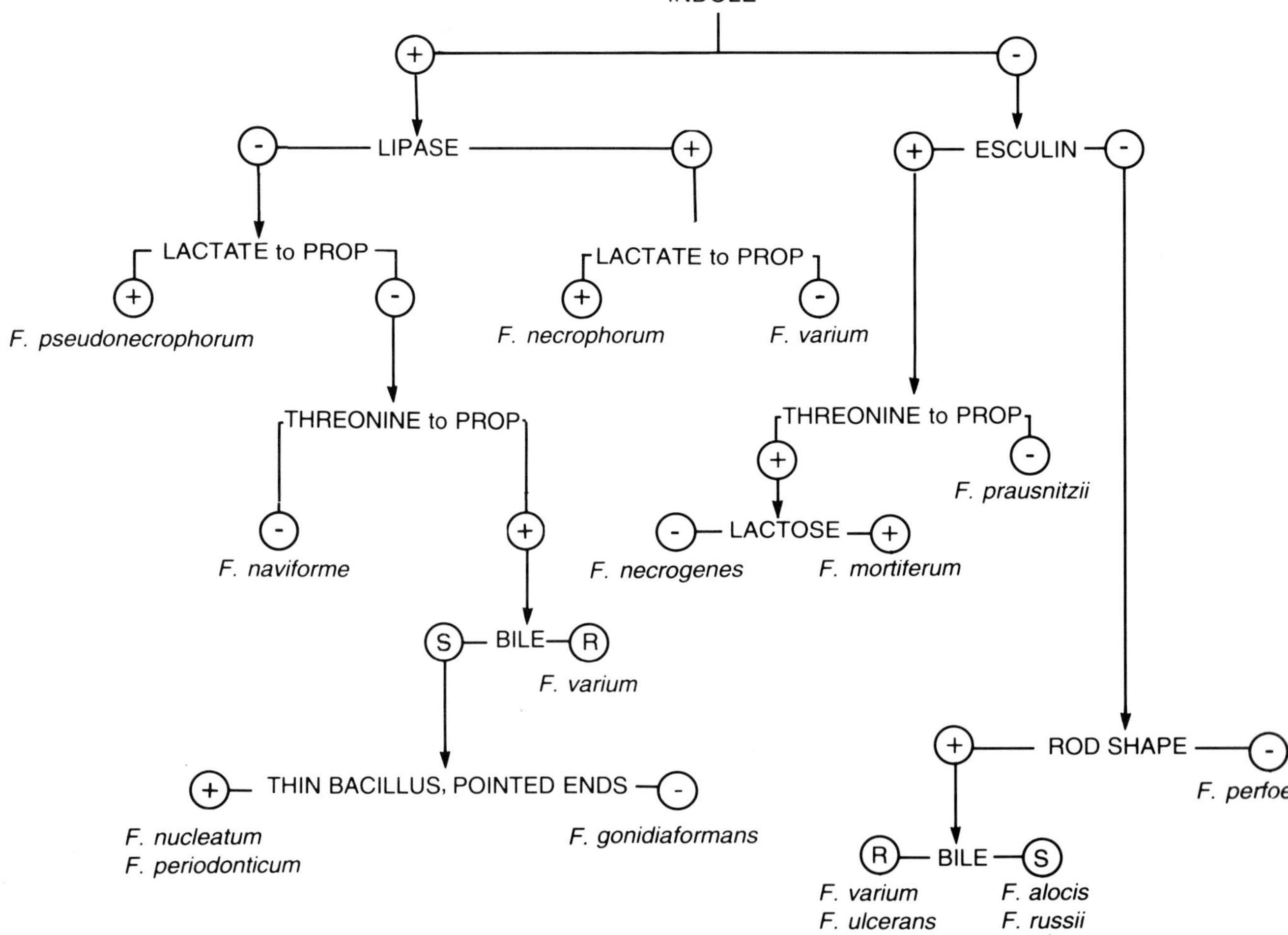

Figure 5-5. Flow chart illustrating scheme for identification of *Fusobacterium* sp.

Table 5-6. Characteristics of *Fusobacterium* species

	DISTINCTIVE CELLULAR MORPHOLOGY	INDOLE	GROWTH IN 20% BILE	LIPASE	ESCULIN HYDROLYSIS	FERMENTATION OF		PROPIONATE FROM	
						LACTOSE (ONPG)	FRUCTOSE	LACTATE	THREONINE
F. gonidiaformans	gonidia forms	+	-	-	-	-	-	-	+
F. mortiferum	bizarre, round bodies	-	+	-	+	+	+w	-	+
F. naviforme	boat shape	+	-	-	-	-	-	-	-
F. necrophorum		+	-+	+	-	-	-w	+	+
F. nucleatum[1]	slender, pointed ends	+	-	-	-	-	-w	-	+
F. pseudonecrophorum		+	-+	-	-	-	-w	+	+
F. russii[2]		-	-	-	-	-	-	-	-
F. ulcerans	round bodies, central swellings	-	+	-	-	-	-	-	+
F. varium		V	+	V[3]	-	-	W+	-	+

[1] *F. periodonticum* shares the same characteristics.
[2] *F. alocis* and *F. sulci* share the same characteristics.
[3] May be positive after 5 to 7 days incubation

published). A new species, *F. ulcerans,* isolated from tropical ulcer, resembles *F. mortiferum* morphologically, but is esculin- and lactose-negative.[1] The Gram stain will differentiate *F. ulcerans* (very pleomorphic) from lipase-negative strains of *F. varium.* In addition, *F. ulcerans* is fructose-negative, whereas *F. varium* is fructose-positive. We have found that *F. periodonticum,*[213] an oral isolate, is biochemically almost indistinguishable from *F. nucleatum.* Fermentation of galactose can be a useful test in differentiating these two species: *F. nucleatum* is negative and *F. periodonticum* is positive. Two other fusobacteria recently described from the gingival sulcus, *F. alocis* and *F. sulci,*[41] are indole-negative and thus easily distinguished from *F. nucleatum,* but indistinguishable phenotypically from each other and *F. russii. Capnocytophaga* species, capnophilic bacteria found as part of the indigenous oral flora, are morphologically similar to fusobacteria on Gram stain (Figure 5-6).

Other gram-negative non-motile bacteria listed in Table 5-1, that were not discussed here, are mainly animal or environmental strains. Two pectinolytic bacteria, *B. pectinophilus* and *B. galacturonicus,*[119] as well as *Megamonas hypermegas,*[203] are found in human feces, but have not been described from clinical specimens.

Gram-negative cocci

Veillonella, Acidaminococcus, and *Megasphaera* comprise the genera of anaerobic gram-negative cocci. *Veillonella* sp. are part of the indigenous mouth, upper respiratory, gastrointestinal tract and vaginal flora; *Megasphaera* and *Acidaminococcus* are part of the intestinal flora. *Veillonella* is isolated more frequently from clinical specimens than the other gram-negative cocci. Table 5-7 contains a key for differentiating these three genera.

Gram-positive cocci

The anaerobic gram-positive cocci are part of the indigenous flora of humans (Table 1-1) and are frequently isolated from clinical specimens (Table 1-2). The genera of clinical importance are *Peptostreptococcus, Streptococcus,* and *Gemella* (a recent taxonomic study has transferred *Streptococcus morbillorum* to the genus *Gemella*).[136] Other anaerobic cocci, *Coprococcus, Peptococcus, Ruminococcus, Sarcina,* and *Staphylococcus saccharolyticus,* are rarely isolated from clinical specimens. Table 5-8 (a and b) and Figure 5-7 show differential characteristics for the gram-positive cocci.

Gas-liquid chromatography and biochemical tests are needed for genus and species level identification of most anaerobic gram-positive cocci except for *P. anaerobius* (SPS-

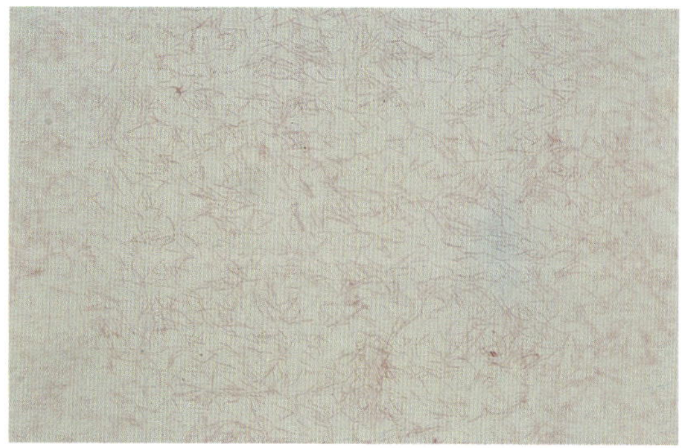

Figure 5-6. Gram stain of *Capnocytophaga* sp., a capnophilic organism that can be confused with *Fusobacterium nucleatum* due to similar cellular morphology.

Table. 5-7. Characteristics of anaerobic gram-negative cocci

	NITRATE REDUCTION	CATALASE	GLUCOSE	FATTY ACIDS FROM PYG
Veillonella sp.	+	V	-	A P
Acidaminococcus fermentans	-	-	-	A B
Megasphaera elsdenii	-	-	+	a ib b iv v C

susceptible), *P. asaccharolyticus* (indole-positive and alkaline phosphatase-negative), and *P. hydrogenalis* (indole-positive and alkaline phosphatase-positive).[235] *P. indolicus,* another indole-positive gram-positive coccus, is rarely isolated from human specimens. The indole-negative, butyric acid producers *P. tetradius* and *P. prevotii* may be differentiated by the fact that *P. tetradius* is strongly saccharolytic, urease-positive, alkaline phosphatase (Rosco disk ALK PHOS)-negative, and β-glucuronidase (PGUA)-positive, whereas *P. prevotii* is weakly saccharolytic (pH 5.7 to 5.9), usually urease-negative, ALK PHOS-positive, and PGUA-negative. *P. anaerobius* is the only coccus that produces isocaproic acid. *P. magnus* and *P. micros* are differentiated by ALK PHOS and cell size: *P. magnus* is ALK PHOS-negative and cells are > 0.6 μm (Figure 4-26) and *P. micros* is ALK PHOS-positive and cells are < 0.6 μm in diameter (Figure 4-25). *P. micros* usually forms short chains and its colonies are small and white and sometimes have a halo of discoloration (slight β-hemolysis) surrounding them.

The microaerophilic streptococci *S. intermedius, S. constellatus,* and *G. morbillorum* may appear as anaerobic organisms on primary culture, are strongly saccharolytic and produce lactic acid as the sole major end-product. Cellular morphology varies in size and form (chains and pairs). Unlike true anaerobes, these organisms are resistant to metronidazole [250] and may become aerotolerant after subculturing. *S. intermedius, S. constellatus,* and *S. anginosus* belong to a DNA homology group formerly known as *S. anginosus* by American workers and "*S. milleri*" by British workers. This group of cocci has been associated with infections and abscess production. A recent study differentiates the species based on specialized enzymatic and biochemical tests.[255] *S. anginosus* is unlikely to be identified using anaerobic methods, since most isolates will grow well on the CO_2 aerotolerance plate. *S. intermedius* and *S. constellatus,* however, may require several subcultures to grow in a non-anaerobic atmosphere and thus would be identified initially as anaerobes (Table 5-8b). Whether the PRAS system results correlate with the newest taxonomic designations has not been determined; the taxonomy of these organisms is still "in flux."

The true obligately anaerobic streptococci (*S. parvulus, S. pleomorphus, S. hansenii*) are rarely or never isolated from clinical specimens. While the origin of *S. parvulus* is unknown,[111] *S. pleomorphus* and *S. hansenii* have been isolated from human feces. One study analyzing the ribosomal RNA of *S. hansenii* and *S. pleomorphus* reported closer

Table 5-8a. Characteristics of *Peptostreptococcus* sp. [1]

	INDOLE	BUTYRIC ACID PRODUCTION	ISOCAPROIC PRODUCTION	GLUCOSE	CELLOBIOSE	LACTOSE	MALTOSE	SUCROSE	UREASE	ALKALINE PHOS	CELL SIZE $\geq 0.6\ \mu m$	FATTY ACIDS FROM PYG
Peptostreptococcus asaccharolyticus	+	+	-	-	-	-	-	-	-	-		B (1 p)
P. hydrogenalis	+	+	-	+	-		-	-		+		B A
P. prevotii	-	+	-	₋w	-	-	W⁻	W⁻	₋+	₋+		B (A 1 p)
P. tetradius	-	+	-	+	-	-	+	+	+	-		B L (a p)
P. anaerobius	-	+	+	W⁻	-	-	W⁻	W⁻	-	-		A IC (ib b iv)
P. productus	-	-	-	+	+	+	+	+	-	-		A l s
P. micros	-	-	-	-	-	-	-	-	-	+	-	A
P. magnus	-	-	-	-	-	-	-	-	-	-	+	A

[1] Recently named species *P. vaginalis*, *P. lacrimalis* and *P. lactolyticus* are not included. (IJSB 1992, 42:602-605)

Table 5-8b. Characteristics of anaerobic and microaerobic *Streptococcus* and *Gemella* sp.

	OXYGEN TOLERANCE	GLUCOSE	CELLOBIOSE	FRUCTOSE	LACTOSE	FATTY ACIDS FROM PYG
Streptococcus parvulus	-	+	+	+	+	L (a)
S. pleomorphus	-	+	-	+	-	L (a)
S. hansenii	-	+	-	-	+	L (a)
Gemella morbillorum	+	+	-	-	-	L (a)
Streptococcus constellatus	+	+	+ᵂ	+	-	L (a)
S. intermedius	+	+	+ᵂ	+	+	L (a)

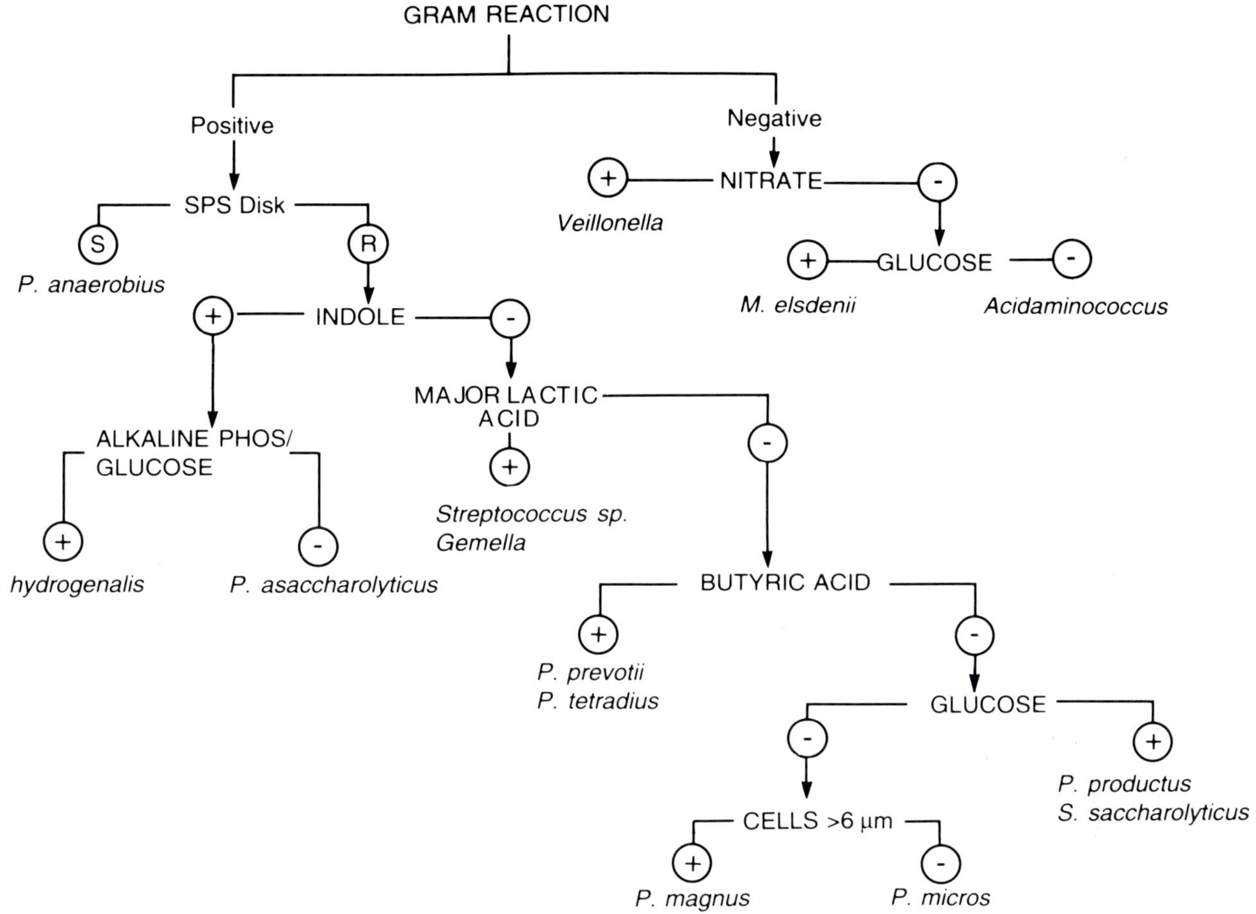

Figure 5-7.   Flow chart illustrating scheme for identification of anaerobic cocci.

similarity to the genus *Clostridium* than to the genus *Streptococcus.*[157]   Further studies are required to establish the validity of these organisms as anaerobic streptococci.

Organisms that do not fit the described species are designated as microaerobic or anaerobic *Streptococcus* sp. or *Peptostreptococcus* sp.

Gram-positive bacilli

Clostridia.   *Clostridium* sp. form a heterogeneous group of anaerobic sporeforming gram-positive bacilli some of which stain gram-variable. They are widely distributed in the environment and constitute part of the indigenous human intestinal flora. Most infections with these organisms are endogenous in origin, although a few noteworthy exogenous intoxications such as food poisoning (*C. perfringens*), botulism (*C. botulinum*) and tetanus (*C. tetani*) occur.

Including the group of clostridia characterized as Nagler-positive (Table 5-9), the *Clostridium* sp. of clinical importance may be subdivided into three groups based on having proteolytic (gelatin hydrolysis), saccharolytic (glucose fermentation), or both proteolytic and saccharolytic properties (Table 5-10). The organisms listed in Tables 5-9 and 5-10

Table 5-9. Characteristics of Nagler-positive *Clostridium* species.

	INDOLE	GELATIN HYDROLYSIS	UREASE	MOTILITY	GROWTH ENHANCED BY 1% MANNOSE	REVERSE-CAMP TEST
C. perfringens	-	+	-	-		+
C. baratii	-	-	-	+		-
C. bifermentans	+	+	-	+	+	-
C. sordellii	+	+	+⁻	+	-	-

represent the more commonly isolated pathogenic clostridia. Potentially clinically significant organisms such as *C. butyricum, C. clostridioforme, C. innocuum,* and *C. ramosum* can be resistant to clindamycin and multiple cephalosporins. There are numerous other species not listed, and these may be identified using Bergey's Manual or the VPI Anaerobe Laboratory Manual and current literature.[39,109,111]

Certain factors must be considered when identifying *Clostridium sp.* First, it is important to note that some of the clostridia destain regularly and appear gram-negative; however, the special potency antibiotic disk pattern (colistin-resistant and vancomycin-sensitive) will demonstrate the gram-positivity of the isolate. Second, the spores of some species are rarely detected microscopically, so that an ethanol or heat spore test may be necessary. Even by these means, however, it may sometimes be difficult to detect spore formation. Third, the colony morphology of pure cultures may be variable, so that the culture appears mixed. Subcultures of single colonies yield the same variable types. Fourth, the aerotolerant clostridia may be confused with *Bacillus* or *Lactobacillus* sp. *Clostridium* species sporulate anaerobically only, grow much better anaerobically (larger colonies), and are almost always catalase-negative, whereas *Bacillus* sp. sporulate aerobically only, usually grow better aerobically, and are usually catalase-positive. Aerobically grown *C. tertium* shares similar colonial and cellular morphology with *Lactobacillus* species (Figures 5-8 and 5-9).

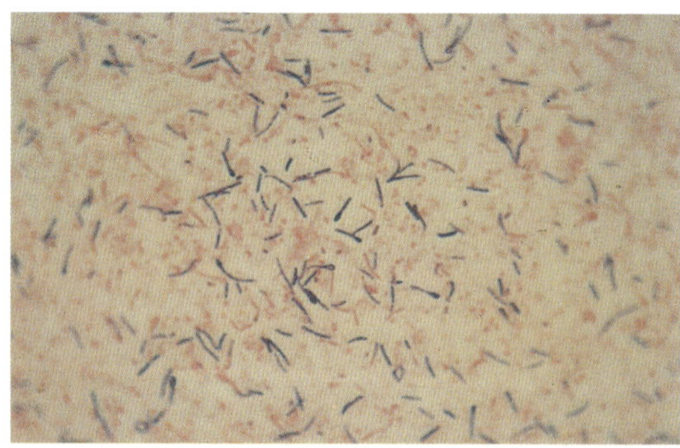

Figure 5-8. Gram stain of *Clostridium tertium* grown anaerobically; note presence of spores.

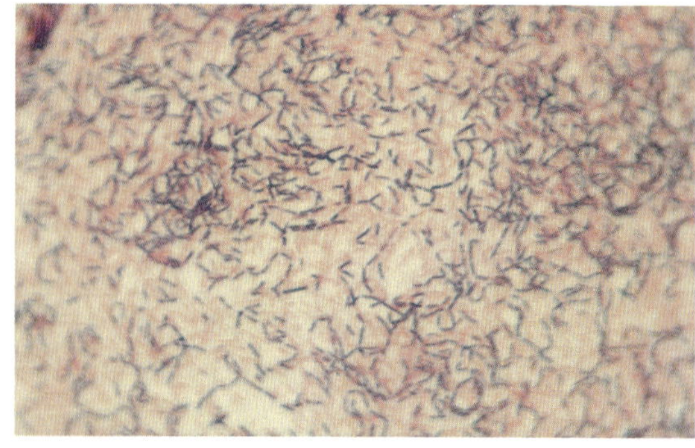

Figure 5-9. Gram stain of *Clostridium tertium* grown aerobically; note absence of spores and lactobacillus-like morphology.

Three of the Nagler-positive clostridia are discussed in the Level II identification section; their characteristics are summarized in Table 5-9. The rare urease-negative strains of *C. sordellii* are difficult to differentiate from *C. bifermentans*. *C. bifermentans* does not produce phenylacetic acid (PAA), its growth is enhanced by 1% mannose and it is glutamic acid decarboxylase-negative (only a few strains have been tested); whereas *C. sordellii* may produce PAA, its growth is not enhanced by 1% mannose and it is glutamic acid decarboxylase-positive.[121]

C. novyi type A is lecithinase- and lipase-positive and may swarm. It is infrequently isolated. Types B, C, and D vary in these properties and are unlikely to be isolated from human clinical material. *C. sporogenes* is lipase-positive, swarms, and has colonies that firmly adhere to the agar. It is phenotypically the same as *C. botulinum* types A, B, and F. Toxin neutralization in mice, polyacrylamide gel electrophoresis (PAGE) of soluble cellular proteins, or GLC of trimethylsilyl derivatives of whole cell hydrolysates are necessary for differentiating *C. sporogenes* and *C. botulinum*. The different *C. botulinum* types vary in saccharolytic, proteolytic, and lipase activity. Suspected *C. botulinum* isolates or possible *C. botulinum*-containing material (e.g., suspected food in food poisoning cases) should be referred to the local or state public health laboratories. Even minute amounts of toxin can be fatal; great care should be exercised when handling suspicious material. The swarmer *C. septicum* produces a Medusa head-like colony in 4 to 8 hours (Figure 5-10) that becomes a heavy film of growth that covers the plate by 24 to 48 hours. The isolation of *C. difficile* from stool samples of patients with suspected antimicrobial-associated diarrhea or pseudomembranous colitis is discussed in Chapter 6. This organism produces a characteristic horse stable odor, forms pleomorphic colonies, and fluoresces chartreuse and produces a characteristic yellow ground-glass colony on cycloserine cefoxitin fructose agar (CCFA).[85]

C. ramosum is a thin, gram-variable rod that has a small round or oval terminal spore, when present (Figure 5-11). It is one of the *Clostridium* sp. that may be resistant to clindamycin and multiple cephalosporins. *C. clostridioforme* stains gram-negative and forms characteristic elongated football-shaped cells that rarely show spores (Figure 5-12). It may be misidentified initially as a *Bacteroides* sp. or *Fusobacterium* sp. *C. oroticum* has a similar cell shape, but it often has spores and forms chains as seen on the Gram stain smear. *C. symbiosum* has the same cell morphology as *C. oroticum,* but is asaccharolytic.

C. tetani may form a thin film of growth over the entire plate, especially on moist media. This characteristically drumstick-shaped gram-positive bacillus is rarely isolated from clinical material of patients with tetanus.

Gram-positive nonsporeforming bacilli.   The nonsporeforming gram-positive bacilli consist of several genera that are differentiated from each other by their metabolic end-products as described in Table 5-1. (See Chapter 7 for methods and discussion.) These bacteria form part of the indigenous human flora at most sites (Table 1-1) but occur infrequently in infections. Various *Actinomyces* species and *P. propionicum* are the major agents in actinomycosis. *Bifidobacterium dentium* has been isolated from a few serious pulmonary infections. *Propionibacterium* sp. are involved occasionally in infections (usually endocarditis and infections related to implanted devices such as ventriculo-atrial shunts or artificial hip joints). Species identification of most of the gram-positive nonsporeforming bacilli is beyond the scope of this manual and of clinical need, except for *Actinomyces* and certain other rare species.

Table 5-10. Characteristics of gram-positive sporeforming bacilli [1]

	GELATIN HYDROLYSIS	GLUCOSE FERMENTATION	LECITHINASE	LIPASE	INDOLE	BUTYRIC ACID IN PYG	ISOACIDS IN PYG	AEROBIC GROWTH	UREASE	MILK REACTION	LACTOSE	MALTOSE	FRUCTOSE	CELLOBIOSE	ARABINOSE	MANNOSE	XYLOSE	NITRATE REDUCTION	SPORE LOCATION	FATTY ACIDS FROM PYG
													FERMENTATION OF							
Saccharolytic proteolytic																				
C. bifermentans	+	+	+	-	+	(+)	(+)	-	-	d	-	W^-	V	-	-	$-^w$	-	-	OS	A (p ib b iv ic l s)
C. sordellii	+	+	+	-	+	(+)	(+)	-	$+^-$	d	-	W^+	V	-	-	$-^w$	-	-	OS	A (p ib b iv ic l)
C. perfringens	+	+	+	-	-	+	-	-		d^c	+	+	+	$-^+$	-	+	-	V	OS^2	A B L (p s)
C. novyi type A	+	+	+	+	-	+	-	-		c	-	V	$-^w$	-	-	-	-	-	OS	A P B
C. sporogenes	+	+	-	+	-	+	+	-		d	-	$-^w$	$-^w$	-	-	-	-	-	OS	A B ib iv (p v ic l s)
C. cadaveris	+	+	-	-	+	+	-	-		d^c	-	-	V	-	-	$-^w$	-	-	OT	A B (l s)
C. septicum	+	+	-	-	-	+	-	-		cd	+	+	+	$+^w$	-	+	-	V	OS	A B (p l)
C. difficile	+	+	-	-	-	+	+	-		-	-	-	+	$-^w$	-	W^-	$-^w$	-	OS	A ib B iv IC (v l)
C. putrificum	+	+	-	-	-	+	+	-		d	-	$-^w$	$-^w$	-	-	-	-	-	OT	A ib B iv (p v ic l s)
Saccharolytic non-proteolytic																				
C. baratii	-	+	+	-	-	+	-	-		c	$+^w$	$+^w$	+	+	-	+	-	$+^-$	RS	A B L (p s)
C. tertium	-	+	-	-	-	+	-	+		c	+	+	+	+	-	+	V	$+^-$	OT	A B L (s)
C. butyricum	-	+	-	-	-	+	-	-		c	+	+	+	+	$+^-$	+	+	-	OS	A B (l s)

Table 5-10. Characteristics of gram-positive sporeforming bacilli[1]

	GELATIN HYDROLYSIS	GLUCOSE FERMENTATION	LECITHINASE	LIPASE	INDOLE	BUTYRIC ACID IN PYG	ISOACIDS IN PYG	AEROBIC GROWTH	UREASE	MILK REACTION	LACTOSE	MALTOSE	FRUCTOSE	CELLOBIOSE	ARABINOSE	MANNOSE	XYLOSE	NITRATE REDUCTION	SPORE LOCATION	FATTY ACIDS FROM PYG
													FERMENTATION OF							
C. innocuum	-	+	-	-	-	+	-	-		-	-	-	+	+	-	+	$-^w$	-	OT	A B L (s)
C. ramosum	-	+	-	-	-	-	-	-		c	+	+	+	+	-	+	$-^w$	-	OT[2]	A l (s)
C. clostridioforme	-	+	-	-	$-^+$	-	-	-		c	V	$+^w$	+	V	V	$+^w$	+	$+^-$	OS[2]	A (l s)
C. sphenoides	-	+	-	-	+	-	-	-		c	W^+	+	+	+	$-^w$	$+^w$	V	$+^-$	OS	A (l s)
C. paraputrificum	-	+	-	-	-	+	-	-		c	+	+	+	+	-	+	-	$-^+$	OT	A B L (s)
Asaccharolytic proteolytic																				
C. tetani	+	-	-	-	$+^-$	+	-	-		d^-	-	-	-	-	-	-	-	-	RT	A p B (l s)
C. hastiforme	+	-	-	-	-	+	+	-		d^c	-	-	-	-	-	-	-	$-^+$	S	A B iv ib (p ic)
C. subterminale	+	-	$-^+$	-	-	+	+	-		d^c	-	-	-	-	-	-	-	-	OS	A B ib IV (p ic l s)
C. histolyticum	+	-	-	-	-	-	-	$+^-$		d	-	-	-	-	-	-	-	-	OS	A (l s)
C. limosum	+	-	+	-	-	-	-	-		d	-	-	-	-	-	-	-	-	OS	A (l s)

[1] *C. botulinum* types vary in proteolytic, saccharolytic, and lipase reactions. Send suspected isolates or suspected *C. botulinum* containing material to the appropriate local or state public health agency.

[2] Spores rarely observed.

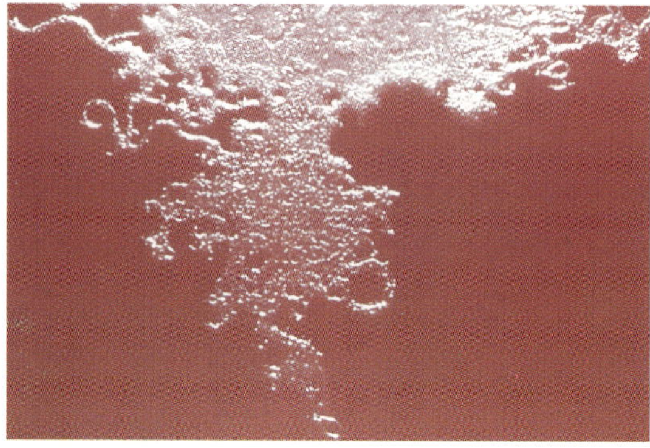

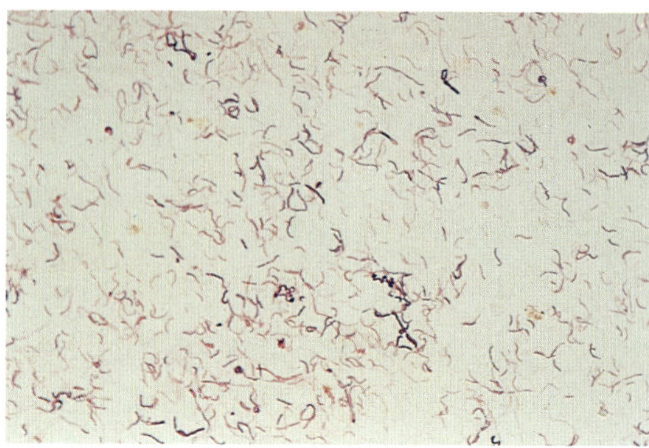

Figure 5-10. *Clostridium septicum* on blood agar after 8 hours; note Medusa-head colony morphology.

Figure 5-11. Gram stain of *C. ramosum*; thin, gram-variable bacilli with noticable spores.

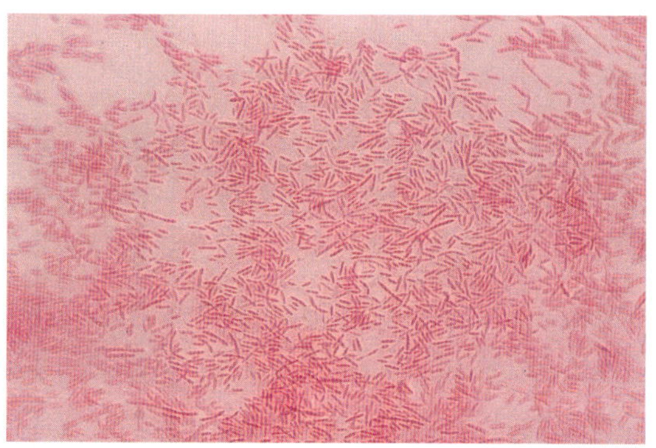

Figure 5-12. Gram stain of *C. clostridioforme*; note gram negatively staining cigar-shaped cells.

Table 5-11 and Figure 5-13 show characteristics used in identifying the more commonly isolated or clinically important bacteria in this group. Several generalizations can be made using results from the nitrate, catalase, and indole reactions: 1) a catalase-positive organism is probably a *Propionibacterium* sp. or *A. viscosus;* 2) a nitrate-positive organism is probably not a *Bifidobacterium* or *Lactobacillus* sp; and 3) an indole-positive organism is probably a *Propionibacterium.* In addition, many strains of *Actinomyces* are microaerobic (CO_2 is required for maximum growth) and are slow growers, requiring more than 48 hours incubation for growth to appear on the primary culture. Branching, beading and diphtheroid-shaped gram-positive bacilli are seen on the Gram stain smear (Figure 5-14). *A. israelii* is noted for its molar tooth colony (Figure 5-15); however, its colonies may also be smooth. Although it is reported that the red pigment production of *A. odontolyticus* is enhanced by exposure to air, these red colonies are also seen after anaerobic incubation (see Figure 4-6). *A. viscosus* is the only catalase-positive *Actinomyces*. *A. meyeri* is a small bacillus and is the only strict anaerobe in this genus. Recent genetic studies have introduced two new species from pre-existing members of *Actinomyces* groups found in human periodontal flora.[124] *Actinomyces* DO8 has been renamed *A. georgiae* and *A. israelii* serotype *II* has been reclassified as *A. gerencseriae* due to low DNA-DNA relatedness with *A. israelii*. Table 5-11 lists the biochemical information necessary for their identification.

Propionibacterium are pleomorphic bacilli that may appear to branch. *P. acnes* is the most commonly isolated species and is often a contaminant in blood cultures. *Arachnia propionica,* found in human actinomycosis and lacrimal canaliculitis, has been renamed

Propionibacterium propionicum.[42] *Bifidobacterium* sp., with bifurcated ends, may also appear to be branching (Figure 5-16); however, these bacilli are generally larger in diameter than *Actinomyces* or *Propionibacterium. Lactobacillus* sp. have straight sides and may occur in chains. The *Eubacterium* genus is comprised of those gram-positive nonsporeforming bacilli that have a GLC pattern not characteristic of the other genera. *E. lentum* is the most commonly isolated species. Its features are described in the Level II species identification section.

Level III identification systems

Several identification methods are currently available, including methods utilizing PRAS biochemicals, many rapid biochemical kits, the Presumpto plate system, and a method based on whole cell fatty-acid analysis. The PRAS biochemical system is still considered the standard method.

PRAS biochemicals are available commercially (Carr-Scarborough, Adams Scientific) or they can be prepared according to the methods published in the VPI Anaerobe Laboratory Manual.[109] CDC utilizes aerobically prepared thioglycolate-based media with bromthymol blue indicator.[58] Table 5-12 lists appropriate biochemicals needed to identify different groups of organisms. Inoculation is done under anaerobic conditions in an anaerobic chamber, or using a special apparatus that allows continuous gas flow into the tubes during manipulation. Another technique utilizes rubber stoppers and the inoculation can be done using a needle and syringe; this has been further modified by replasing the rubber stopper with a Hungate-type screw-cap (PRAS II, Adams Scientific).[20] Inoculation and interpretation procedures are outlined in Appendix B. A short PRAS biochemical scheme for identification of clinical isolates of bile-resistant *Bac-*

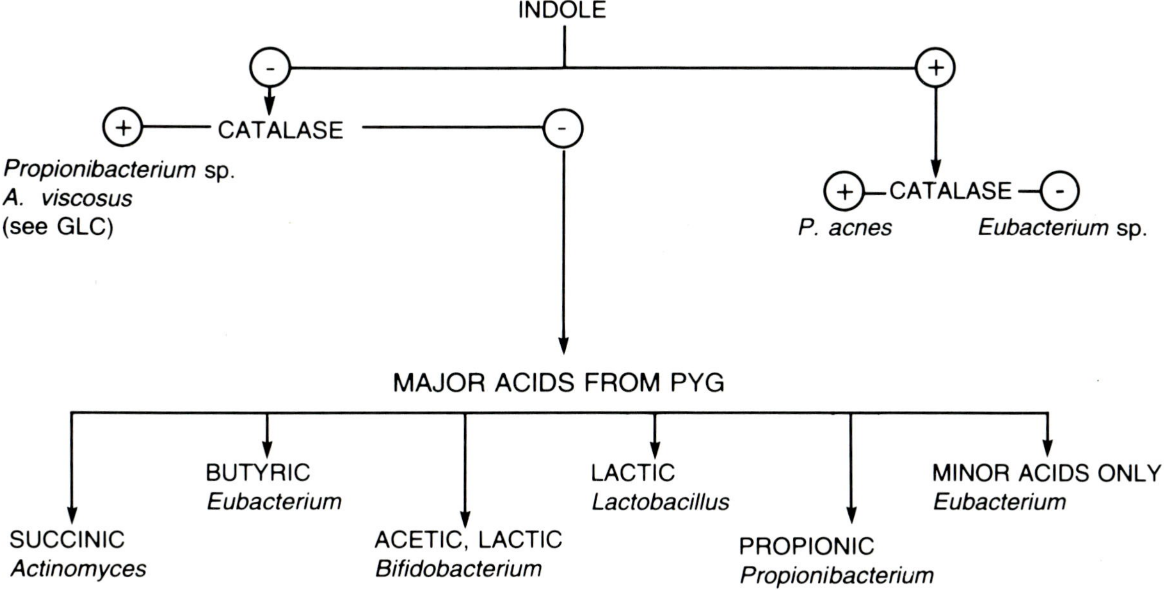

Figure 5-13. Flow chart illustrating scheme for identification of gram-positive nonsporeforming bacilli.

Table 5-11. Level III identification of gram-positive nonsporeforming bacilli

	NITRATE	CATALASE	INDOLE	ESCULIN HYDROLYSIS	GELATIN HYDROLYSIS	UREASE	COLONY RED PIGMENTED	OXYGEN TOLERANCE	GLUCOSE	ARABINOSE	MANNOSE	RAFFINOSE	TREHALOSE	FATTY ACIDS FROM PYG
									FERMENTATION OF					
Actinomyces sp.	+	$-^+$	-	$+^-$		V			+					A S (L)
A. israelii	$+^-$	-	-	+		-	-	AM	+	+	+	+	$+^-$	A L S
A. odontolyticus	$+^-$	-	-	$+^-$		-	$+^-$	AM	+	-	-	-	-	A S
A. naeslundii	+	-	-	+		$+^-$	-	MF	+		-	+	$+^-$	A L S
A. viscosus	$+^-$	+	-	+		$+^-$	-	AM	+		-	+	V	A L S
A. meyeri	-	-	-	$-^+$		-	-	AM	+		-	-	-	A S
A. georgiae	$-^+$	-	-	$+^-$		-	-	F	+		$-^+$	$-^+$	+	A S (l)
A. gerencseriae	$-^+$	-	-	+		-	-	AM	+	-	$+^-$	+	+	A S
Propionibacterium sp.	V	V	$-^+$	V	V			AM	+					A P
P. acnes	+	+	$+^-$	-	+			AM	+					A P (iv L s)
P. granulosum	-	+	-	-	-			AM	+					A P (iv s)
P. avidum	-	+	-	+	+			AM	$+^w$					A P (iv s)
P. propionicus	+	-	-	-	V			AF	+					A P (l s)
Eubacterium sp.	V	-	$-^+$	V				A	$+^-$					
E. nodatum[1]	-	-	-	-				A	-					a B (l s)
E. lentum	+	$-^+$	-	-				A	-					(a l s)
Lactobacillus sp.	$-^+$	-	-	V				AMF	$+^-$					L (a) [L>a]
Bifidobacterium sp.	-	$-^+$	-	$+^-$				AM	+					A L [A>L]

[1]Morphologically similar to *Actinomyces* sp.

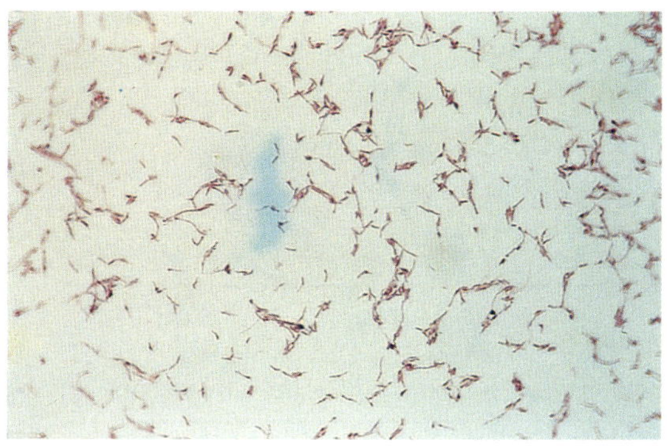

Figure 5-14. Gram stain of *Actinomyces viscosus*; note branching.

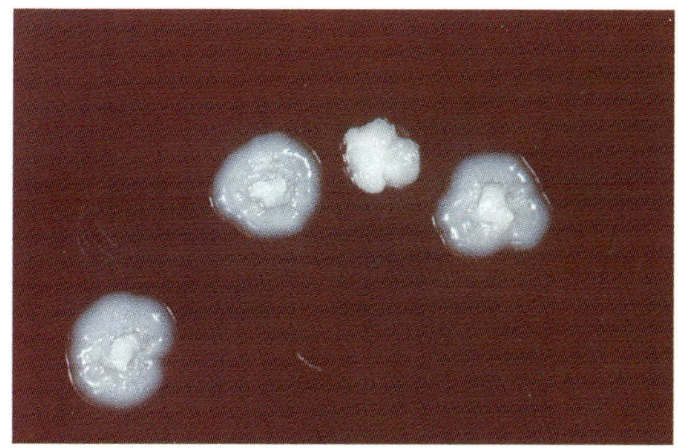

Figure 5-15. *Actinomyces israelii*; note molar-tooth colony.

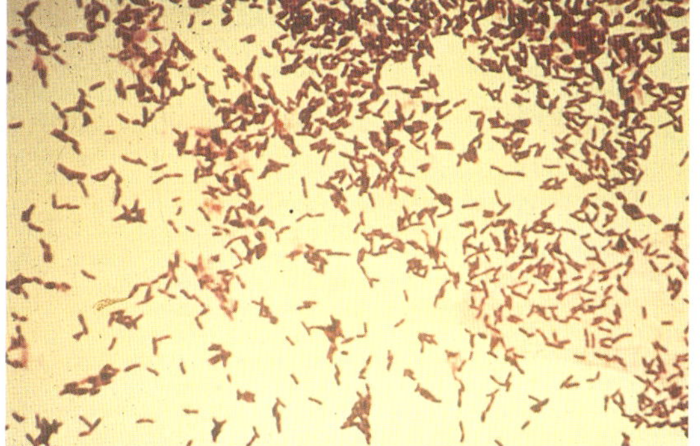

Figure 5-16. Gram stain of *Bifidobacterium* sp.; note pleomorphic bacilli with bifurcated ends.

Table 5-12. Level III PRAS tests for identification of anaerobic isolates

B. fragilis group	Pigmented gram-negative bacillus	Gram-negative bacillus	*Fusobacterium*	Gram-negative coccus
PY	PY	PY	PY	PY
PYG	PYG	PYG	PYG	PYG
ARAB	CB	ARAB	ES	INO$_3$
CB	ES	CB	FRUC	
ES	LAC	FRUC	LAC	
RH	SAL	ES	TH	
SUC	SUC	LAC	LT	
TRE	GEL	RH	INO$_3$	
INO$_3$	INO$_3$	SAL	BILE	
XYLAN (if indole positive)		SUC		
BILE		TRE		
		XYLAN		
		XYL		
		GEL		
		INO$_3$		
		BILE		

Table 5-12. Level III PRAS tests for identification of anaerobic isolates

Gram-positive coccus[1]	Bifidobacterium[1]	Clostridium[1]	Actinomyces / Propioni-bacterium[1]	Gram-positive bacillus; unknown genus[1]
PY	PYG	PY	PY	PY
PYG	ARAB	PYG	PYG	PYG
CB	CB	ES	ARAB	ARAB
ES	FRUC	FRUC	ES	BILE
FRUC	INU	LAC	FRUC	CB
LAC	LAC	ML	INT	ES
ML	ML	MANIT	LAC	FRUC
MANO	MANIT	MANO	ML	INT
SUC	MANO	MELZ	MANIT	INU
TW	MELZ	MEB	RAF	LAC
INO[3]	MEB	RI	RI	ML
	RAF	SAL	SORB	MANIT
	RI	SUC	ST	MANO
	SAL	XYL	TRE	MEB
	SORB	GEL	GEL	MELZ
	ST	MILK	TW	RAF
	TRE	TW	INO[3]	RH
	XYL	INO[3]	UREA	RI
				SAL
				SORB
				SUC
				ST
				TRE
				GEL
				MILK
				TW
				INO[3]
				UREA

[1] Not all biochemicals listed here are shown in the identification tables in this manual. The VPI (109) and Bergey's (111) manuals should be consulted for additional results.

ABBREVIATIONS

PY	Peptone yeast	MEB	PY-melibiose
PYG	PY-glucose	MELZ	PY-melezitose
ARAB	PY-arabinose	MILK	Milk
BILE	PYG-bile	PYR	Pyruvate
CB	PY-cellobiose	RAF	PY-raffinose
ES	PY-esculin	RH	PY-rhamnose
GEL	Gelatin	RI	PY-ribose
FRUC	PY-fructose	SAL	PY-salicin
INT	PY-inositol	SORB	PY-sorbitol
INO₃	Indole-nitrate	SUC	PY-sucrose
INU	PY-inulin	ST	Starch
LT	PY-lactate	TH	PY-threonine
LAC	PY-lactose	TRE	PY-trehalose
LIPASE	Lipase	TW	PY-tween
ML	PY-maltose	UREA	Urea
MANIT	PY-mannitol	XYLAN	PY-xylan
MANO	PY-mannose	XYL	PY-xylose

teroides sp. was recently described by Citron and colleagues.[48]   For further guidelines on identification of particular organisms, refer to the VPI Manual,[109,170,172]   Bergey's Manual,[111] and the CDC Manual.[58]   It should be noted that definitive identification with PRAS biochemicals often requires additional tests such as gas-liquid chromatography and other biochemical tests including catalase, urease, and chromogenic enzyme tests.

The PRAS system is expensive, labor-intensive, and not problem-free. The major problem is interpretation of biochemical results. The following three points may aid in reducing potential errors:

1) The inoculum must be a young culture of at least 2+ turbidity. An old culture contains predominantly non-viable organisms and its use may lead to unreliable results. Since some anaerobes grow poorly in thioglycolate or peptone-yeast broth, the addition of supplements may be necessary to enhance growth. If growth in the thioglycolate broth is less than 2+, perform growth stimulation tests (see Appendix B) and add an appropriate supplement to each PRAS tube. Tween-80 and hemin can be added directly to the inoculum broth.

2) The anaerobe tested must be in pure culture and viable in PRAS biochemical tubes. Do not read biochemical reactions of mixed cultures; reisolate the organism and repeat the biochemical inoculation. If the isolate fails to grow on the purity and viability plate, but there appears to be turbidity in the biochemical tubes, then subculture the PYG to determine that the turbidity is growth of the isolate and not

merely inoculum or a contaminant. If the organism grows on the plate, but not in the biochemical tubes after 2 to 3 days of incubation, then perform growth stimulation tests and reinoculate. Do not incubate non-turbid biochemicals for >5 days, as growth rarely occurs beyond that time.

3) Some anaerobes characteristically grow with heavy turbidity while others do not. Any inconsistencies in growth throughout the series of tubes should be investigated. Unusually heavy growth in one or two tubes suggests either the presence of a growth factor or contamination. Gram stain or subculture an aliquot of material from the questionable tubes to rule out contamination. Poor or no growth in one or two tubes of a series showing good growth may be due to an oxidized tube (pink color due to resazurin indicator change) or a tube skipped during inoculation. *Propionibacterium acnes* and *Staphylococcus* sp. are the usual contaminants. If the indole and nitrate tests are both positive, suspect that *P. acnes* is present in the tubes. An aerotolerance test performed along with the purity and the viability check will help to determine the presence of an aerobic contaminant.

The API 20A (Analytab Products, Inc.) and Minitek (BBL) are microtube biochemical systems. The API 20A strip consists of 16 cupules containing carbohydrates and tests for indole, urease, gelatin and esculin hydrolysis, and catalase. The Minitek system utilizes disks impregnated with carbohydrates as well as substrates for indole, esculin, nitrate, and urease tests. The disks are dispensed into wells of a microtiter tray and a broth suspension of the organism is added to the wells. After 24 to 48 hours of incubation, the color reactions of the API 20A and Minitek systems are read manually and scored. A code number is generated and interpreted from the code book or computerized data base provided by the manufacturer. The color reactions in both of these systems are not always clear-cut, as shades of brown or no color (API 20A) and yellow-orange (Minitek) can make interpretation of the test results difficult. API 20A and Minitek are best suited for identification of saccharolytic, fast-growing organisms, such as the *B. fragilis* group and many clostridia.[97,103,104,115,133,226] Most asaccharolytic organisms cannot be identified and some fastidious organisms fail to grow in the systems (e.g., some *Prevotella* sp.). Even for many saccharolytic organisms, supplemental tests, including bile and GLC, are often required for definitive identification.

Several rapid (4-hour) identification systems based on detection of preformed enzymes have been developed and subsequently evaluated.[4,40,56,97,104,115,133,137,181,198,224,226,238,245] These systems include RapID ANA II (Innovative Diagnostic Systems, Inc.), Rapid ID 32A (BioMerieux, France), ANI Card (Vitek Systems), AN-Ident and API ZYM (Analytab), and MicroScan (American MicroScan). The systems utilize chromogenic substrates for bacterial glycosidases, aminopeptidases and some other enzymes. The inoculum for these systems consists of a bacterial suspension equal to McFarland standard no. 3 to 5. The test organism must be a fresh (24 to 72 hour) subculture on an agar medium that does not contain carbohydrates. Panels are incubated aerobically. The short incubation time and heavy inoculum minimize the risk of a contaminant interfering with the results, which may be a problem with methods based on bacterial growth. After incubation, appropriate reagents are added, and reactions are interpreted either manually or with an automated reader (MicroScan [AutoScan-4], Rapid ID 32A). A specific code number is generated and compared with listings in a code book for identification. The accuracy of these systems can be further increased by certain simple supplemental tests and by GLC. Some systems provide computer-assisted identification (Vitek, Micro-Scan, and Rapid ID 32A). The API ZYM system does not have a data base, but it has been evaluated for its effectiveness for identification of clinically encountered anaerobic bacteria.[104,108,144,163,238] The interpretation of color reactions may cause problems with some of these systems; furthermore, the growth medium used to prepare the inoculum may interfere with the results for some organisms.

Individual chromogenic enzymatic tests also are available commercially (Rosco, Carr Scarborough) and provide convenient, simple tests for identification of anaerobes. A heavy suspension of the organism is prepared in 0.25 ml of saline and incubated at 35°C for 4 to 24 hours aerobically with a tablet containing the substrate. After incubation, reagent is added, if applicable, and the reactions are interpreted.

The Centers for Disease Control (CDC) has developed three four-quadrant plates (Presumpto plates I, II, III) that incorporate test substrates in Lombard-Dowell agar medium.[59,60,152] These plates contain tests for determination of growth in 20% bile; lecithinase and lipase production; esculin, starch, casein and gelatin hydrolysis; production of H_2S and indole; DNase activity; and carbohydrate fermentation (glucose, mannitol, lactose, and rhamnose). Plates are available from Carr Scarborough and Remel Laboratories.

A recently developed method for identifying or assisting in identification of anaerobes is based on analysis of whole cell fatty-acids by capillary column GLC. The method (Hewlett Packard Microbial Identification System [MIS]) is discussed in Chapter 7.

Chapter 6

Laboratory Tests for Diagnosis of Clostridium difficile *Enteric Disease*

GENERAL PRINCIPLES

Antimicrobial-associated colitis is usually caused by overgrowth of *Clostridium difficile* in the bowel. The enterotoxin (toxin A) and the cytotoxin (toxin B) of *C. difficile* are both contributory, with the enterotoxin felt to be more important in the pathogenesis of diarrhea. Definitive diagnosis of pseudomembranous colitis requires visualization of pseudomembrane via endoscopy (and documentation of the microbial cause), which is not often performed. Although none are diagnostic by themselves, laboratory tests can contribute evidence to substantiate a clinical diagnosis.

Tests currently available to the routine laboratory include: 1) culture, 2) latex agglutination for detection of a cell-associated protein, 3) enzyme linked immunosorbent assays for enterotoxin A only or for both enterotoxin A and cytotoxin B, and 4) cell culture assay for cytotoxin, commercial system (Bartels Immunodiagnostics) or conventional method. Several commercial sources are available for the latex agglutination and the ELISA test formats. All three methods test for different characteristics so test results cannot be compared directly. To increase the sensitivity of laboratory tests for *C. difficile,* two different tests may be performed on the same specimen.

Because some patient populations may harbor the organism in stool in the absence of symptoms, it is important that the laboratory select specimens to increase the specificity of the test results. Children less than 1 year old have a relatively high carriage rate. Their stools should not be tested. In addition, we recommend that only unformed stools (stools not colon-shaped) be tested routinely for *C. difficile.* Of course, extenuating circumstances identified by physicians should allow for flexibility in testing of certain specimens.

Cultures are the least specific but most sensitive method to detect possible disease related to *C. difficile.* However, studies have shown that as many as 20% or more of patients in hospitals where this organism is endemic become colonized asymptomatically with the organism, particularly if they are receiving certain antimicrobial agents.[178] Culture results are usually not available for 48 hours after specimen receipt. Thus, culture is suitable for prevalence studies and research projects, but is not practical for clinical purposes.

Alternative broth culture screening procedures using gas-liquid chromatography for detection of para-cresol or other end-products (such as the presence of butyric, isovaleric, valeric, and isocaproic acid) have been used.[126] The addition of 0.1% sodium taurocholate to broth may enhance spore germination and increase yield of cultures.[262] Add cefoxitin (16 mg/L) and sodium taurocholate (0.1%) to commercially purchased PRAS broths (PY fructose or chopped meat) to prepare a suitable enrichment broth for epidemiologic studies. Cycloserine-cefoxitin fructose agar (CCFA), a selective agar developed by George and colleagues [85] is recommended for plate cultures. The addition of taurocholate or mannitol has not resulted in better yields with agar.[117]

Additional tests are in the development stage. A DNA probe for toxin A has been used for identification of isolates.[264] It does not, however, perform reliably with stool specimens at this time. The polymerase chain reaction (PCR) has been used to amplify the enterotoxin A gene of *Clostridium difficile* and may prove useful for direct detection in specimens.[134] However, because the organism and its toxins may be present in asymptomatic individuals, laboratory tests for its presence will never constitute the sole diagnostic modality.

LABORATORY PROCEDURES

Specimen collection for all procedures

Collect stool in a leakproof, tightly lidded container. Do not contaminate with urine. Deliver promptly to laboratory or freeze at -20°C or colder if unable to process within 2 days of collection.

Culture procedures

Principle. Stool specimens from patients with diarrhea can be tested for the presence of *Clostridium difficile* by culturing on a selective medium. The antibiotics cycloserine and cefoxitin inhibit the growth of most bacteria other than *Clostridium difficile*. Fructose is metabolized by *C. difficile* more actively than is glucose, and neutral red acts as an indicator to detect proteolysis of the medium. Colonies of *C. difficile* break down proteins in the medium, which results in production of alkaline end-products that turn neutral red a yellow color. *C. innocuum* is resistant to cefoxitin and may grow on CCFA, although the colony and cellular morphology are different than those of *C. difficile*. Either fresh (less than 2 hours from collection) or frozen stool is acceptable, but fresh stool yields higher counts. Rectal swabs and frozen stool should be enriched in a broth medium to encourage spore germination before plating.

Note that not all isolates of *Clostridium difficile* are toxin-producing, and that even the presence of toxin-producing *Clostridium difficile* may not be contributing to disease in a particular host.

Specimen evaluation.   Stool specimens less than 2 hours old or frozen at -20°C are acceptable for culture. Rectal swabs are not recommended except for broth enrichment epidemiologic studies.

CCFA medium.   Prepare the medium according to the directions in Appendix C. Store the plates in plastic bags at 2-8°C for up to 8 weeks.[65] CCFA plates are available commercially but, as for most anaerobic media, freshly prepared or anaerobically prepared and stored (PRAS) plates are best.

Specimen processing

1.   Mix the stool specimen thoroughly and save a 2 ml aliquot for toxin testing, if desired. If the toxin test will not be performed within 2 days, freeze the aliquot at -20°C or colder.
2.   Perform quantitative cultures for *C. difficile* in an anaerobic chamber. Dilute 1 g or 1 ml of stool in 9 ml of diluent (e.g., 0.05% yeast extract) and vortex in the chamber in tubes containing glass beads until the suspension appears homogeneous (approximately 60 sec). Make serial 10-fold dilutions in diluent and plate duplicate 0.1 ml samples from dilutions 10^{-1}, 10^{-3}, and 10^{-5} onto individual CCFA plates using a rotator-pipet method.[6] After 24-48 h anaerobic incubation at 37°C, count the number of characteristic colonies (see below) on those plates yielding 30-300 CFU. **Note: if plates are incubated in an anaerobic chamber or Bio-Bag, they may be inspected after 24 h; if colonies are large enough to count at this stage, the incubation can be concluded. Jars, on the other hand, must not be opened before 48 hours.** Calculate the numbers of CFU per ml stool by multiplying the average CFU on the plates chosen times all dilution factors (e.g., an average of 30 colonies on the two plates that received 0.1 ml from the 10^{-3} dilution means a count of 3×10^5 CFU/ml stool).
3.   Perform semiquantitative cultures by inoculating approximately 0.1 g or 2 to 3 drops of liquid stool (use a Pasteur pipet) onto the first quadrant of a CCFA plate and streak for isolation. Use the same grading system as described in the text for clinical specimens (1+ to 4+); this roughly correlates with CFU/ml of *C. difficile* in stool (4+ corresponds to $\geq 10^5$ CFU/ml).

Isolation and identification

1.   The distinctive colonies of *C. difficile* on CCFA are approximately 4 mm in diameter (if well isolated), yellow, ground-glass in appearance, circular with a slightly filamentous edge, and low umbonate to flat in profile (see Figure 3-8). The initial orange-pink color of the medium is often changed to yellow for 2-3 mm around the colonies. *C. difficile* produces a "horse stable" odor that is detectable from the primary culture plate, but is easier to detect from a pure culture plate. The colonies fluoresce chartreuse (yellow-green) on CCFA, although nonspecific chartreuse fluorescence by the specimen itself and other anaerobes makes this characteristic less reliable. Other organisms may grow on CCFA, but their colonies are usually smaller and differ morphologically from those of *C. difficile.*
2.   Issue a presumptive preliminary report when colonies morphologically resembling *C. difficile* are seen on CCFA.
3.   Subculture *C. difficile*-like colonies to blood agar or another CCFA plate for purity and to chocolate agar for aerotolerance test (incubate in 5-7% CO2). Perform a Gram stain to detect the presence of spores and verify the morphology. *C. difficile* sporulates readily on blood agar but rarely on CCFA. Inoculate well-isolated colonies into thioglycolate broth for subsequent end-product analysis by GLC.

4. Examine the pure culture plate for colony morphology, fluorescence, and odor. After the thioglycolate broth has incubated for 48 h, it should be processed by GLC as described in Chapter 7.

Final report

1. Report culture positive for *C. difficile* if an organism is isolated that is an anaerobic gram-positive sporeforming bacillus with typical colony morphology, produces "horse stable" odor, fluoresces chartreuse, and produces isocaproic acid among a number of end-products. If determined, report the CFU/ml or CFU/g dry weight of stool. Alternatively, report the semiquantitative culture result.
2. If the above criteria are not met, report the culture as negative for *C. difficile.*
3. Inoculate equivocal isolates into PRAS media for definitive identification.

Quality control. Maintain a viable culture of *Clostridium difficile* in chopped meat broth. Subculture a few drops to each new batch of medium and weekly thereafter and observe for characteristic colonies. Note that a laboratory-acclimatized strain may yield better growth than fresh clinical isolates; thus its growth is only a rough approximation of media acceptability.

Cytotoxin testing using a commercial system

Principle. Production of cytotoxin by fecal isolates of *Clostridium difficile* increases the possibility of involvement in clinical disease. Certain mammalian cells display characteristic cytopathic effect (CPE) in the presence of *C. difficile* toxin that is neutralized by specific antitoxin. Stool from patients with *C. difficile* enteric disease will contain detectable cytotoxin in the filtered supernatant, obviating the need for culture. A commercial product containing human foreskin fibroblast monolayers in a microbroth dilution plate format (Baxter Healthcare Corp., Bartels Diagnostics Division) has been shown to detect cytotoxin adequately.[265] Patient stool filtrates are tested concurrently with neutralized filtrates. Cytopathic effect that is inhibited by antitoxin is assumed to have been produced by *C. difficile* cytotoxin.

Note that the cytotoxin is not thought to play as strong a role in disease pathogenesis as is the enterotoxin. Enterotoxin, however, is produced in much smaller amounts than cytotoxin and is thus more difficult to detect in a bioassay system.

Procedure

1. Follow manufacturer's instructions for preparing fresh or defrosted stool. Briefly, centrifuge stool at 3,000 x g for 30 minutes and remove supernatant. **Note: use aerosol-protective containers in the centrifuge. If possible, place centrifuge in a biological safety cabinet or fume hood.**
2. Filter sterilize supernatant through the membrane filter. **Note: wear gloves and face shield and perform the procedure in a biological safety cabinet. The back pressure on the filter/syringe assembly has been known to cause stool to be expelled with force, creating aerosols which may contain infectious hepatitis, HIV, or other virus.**
3. Dilute in diluent (provided) to produce a 1:20 final concentration. Alternatively, to avoid nonspecific toxicity, perform tests at a final concentration of 1:100 stool filtrate to diluent. Patients with clinically significant amounts of cytotoxin typically

will be positive at this dilution. Remove 0.1 ml of each final dilution to a separate tube or empty microbroth dilution well (for the addition of antitoxin).

4. To the 0.1 ml aliquot of each specimen, add an equal amount of neutralizing antitoxin. Allow the antitoxin mixtures to incubate at room temperature for 30 minutes. During this time, prepare a template indicating how samples are to be allocated to wells.
5. Carefully remove the plastic cover from enough strips of wells (8 wells/strip) to accommodate all patient specimens and controls.
6. Add 0.1 ml of each control (see "Quality control," below), patient specimen filtrate, and neutralized filtrate to a separate well. Carefully replace the cover, being sure not to allow spillover.
7. Incubate the microtiter plate at 35-37°C overnight and observe for CPE (cells round up, become more refractile, and eventually detach from monolayer). Record initial results and reincubate an additional day before reporting final results. See "Interpretation of results," below.
8. Mark the strips that were used (so you won't reuse them accidentally) and return the microtiter plate to the incubator. The remaining unused wells may be used for additional test runs.

Interpretation of results

Specimen well	Neutralized well	Interpretation
CPE positive	Normal monolayer	Cytotoxin present
Normal monolayer	Normal monolayer	Cytotoxin negative
CPE positive	CPE positive	Nonspecific cytotoxicity. Must perform test again with sample diluted (1:10, 1:20, 1:40, 1:80). If able to dilute out CPE in neutralized wells, then cytotoxin test is positive.

Reporting results

1. Report all positive results and preliminary negative results after 24 hours.
2. Reincubate the plate for an additional 24 hours to detect late-appearing cytotoxicity. Report all final negative results at 48 hours.
3. For nonspecific toxicity, report results as highest titer at which nonspecific toxicity still persists and titer at which specific cytotoxicity is detected. If unable to dilute out nonspecific toxicity, report "Nonspecific toxicity; unable to interpret results."
   Note: freezing and thawing the stool filtrate several times may remove nonspecific toxic factors (and reduce the toxin titer).

Quality control

1. As each new microtiter plate is received, examine all wells to insure that the monolayers are confluent and healthy-looking. Before performing each test, again observe the monolayers to be used and verify their suitability.

2. Each time the test is performed, include a positive control (diluted 1:10 in diluent) and a neutralized positive control, an antitoxin control, and a diluent control. The positive control should exhibit CPE throughout the well, which is absent in the neutralized well. No CPE should be visualized in wells containing antitoxin alone and diluent alone.

Latex agglutination test for detection of Clostridium difficile-*associated protein*

Principle.   Antibody directed against the enzyme glutamate dehydrogenase (associated with *C. difficile* and several other organisms) is bound to the surface of polystyrene (latex) particles. When the dehydrogenase antigen reacts with the particles, a macroscopically visible precipitate develops. Presence of a detectable amount of this protein in stool filtrate has been shown to correlate with the presence of *C. difficile* toxin in the specimen.

Procedure

1. Follow manufacturer's instructions (Meridian Diagnostics, Inc.) for preparing fresh or defrosted stool. Briefly, centrifuge stool at 3,500 x g for 15 minutes and remove supernatant. **Note: use safety cups in the centrifuge, wear gloves and face shield, and perform the procedure in a biological safety cabinet.**
2. Place appropriate amount of supernatant onto each of two circles on the slide provided. Add detector latex particles to one drop and control latex particles to the other drop, following instructions for the system.
3. Mix the reagent drops and rotate the slide for the specified time. Observe for agglutination. A positive result shows greater agglutination in the detector latex well than in the control latex well.

Interpretation of results

Detector latex well	Control latex well	Interpretation
Easily visible agglutination	Faint granular or no agglutination	Positive test result
Faint granular or no agglutination	Faint granular or no agglutination	Negative test result
Visible agglutination	Visible agglutination equal to the detector latex	Nonspecific, unable to interpret results

---

Quality control.   Each time the test is performed, test a drop of positive control antigen (supplied) as above. The positive control should show strong agglutination with the detector latex and fine granular or no agglutination with the control latex.

Principle.   The enterotoxin A of *C. difficile,* although produced in lesser amounts than the cytotoxin, is thought to be important for development of symptomatic disease. This ELISA test (Meridian) uses a polyclonal capture antibody and a monoclonal detector antibody conjugated to horseradish peroxidase. If the toxin is present in stool supernatant, it will bind to the polyclonal antibody attached to the inner surface of wells in a microtiter plate. Addition of the monoclonal conjugate and its substrate allow detection of the bound antigen. Other test systems may use slightly different formats.

Procedure

1.   Before testing, be certain that all reagents are at room temperature.
2.   Prepare buffer as described in manufacturer's instructions.
3.   Add sample diluent (supplied) to a small test tube for each specimen to be tested.
4.   Prepare stool for test. Some manufacturers require centrifugation and filtration, others use a simple dilution. **Note: do this in a biological safety cabinet to avoid creating aerosols.**
5.   Remove sufficient microdilution well strips from the package to test all specimens and controls.
6.   Follow specific manufacturer's instructions for handling, inoculating, incubating, washing, and reading plate.
7.   Read spectrophotometrically, if at all possible, or visually if no other method is available. **Note: check manufacturer's instructions; only certain test systems can be interpreted visually. If an ELISA reader is required by the manufacturer, do not try to read tests visually.**

Interpretation of results.   For visual interpretation, any change in color signifies presence of enterotoxin A in the stool supernatant. If the color is so faint that there is a question about its presence, then the result is interpreted as negative. Spectrophotometrically, positive wells must read >0.10 OD at 450 nm or as stated in the product insert.

Quality control.   Each time the test is performed, a positive and a negative control (supplied) must be tested along with patient samples.

Chapter 7

Gas Liquid Chromatography (GLC) for Metabolic End Products and Cellular Fatty Acid Methyl Ester Analysis

APPLICATIONS AND LIMITATIONS OF GLC FOR END PRODUCT ANALYSIS IN ANAEROBIC BACTERIOLOGY

Bacteria growing in the absence of oxygen are unable to oxidize nutrients to carbon dioxide and water. Instead they produce intermediate products consisting of short chain fatty and organic acids and alcohols that are distinctive for the various groups. Because the enzymes responsible for metabolism are genetically stable, the end-products produced provide a fingerprint that is distinctive and useful, in conjunction with other tests, for identification. This is particularly important for the gram-positive bacilli, which otherwise may be difficult to classify to the genus level. Table 5-1 lists end products produced by the genera of anaerobes isolated from clinical specimens and indigenous flora sources.

The gram-positive bacilli can readily be identified to the genus level by their end products. *Propionibacterium* sp. typically produce a major amount of propionic acid; this distinguishes them from other gram-positive bacilli. *Actinomyces* sp. produce succinic acid in addition to acetic and, in some species, lactic acid. Succinic acid production is increased in the presence of CO_2 for some species. Lactobacilli produce lactic acid as their sole major end product of metabolism. *Bifidobacterium* sp. also produce large amounts of lactic acid; however, they produce acetic acid in a ratio of at least 2:1 to the lactic acid, which distinguishes them from lactobacilli. The genus *Eubacterium* is a "catch-all" genus that encompasses all non-sporeforming gram-positive bacilli that have end products other than those described above. The different species within this genus produce a variety of end products which can be helpful for their identification. The genus *Clostridium* is very heterogeneous and consists of over 90 described species of sporeforming bacilli. The genus is not defined by the production of any specific end products. Many of the species are biochemically similar; thus, end product analysis can be useful for their identification.

Among gram-negative bacilli, the genus *Fusobacterium* is defined by the production of butyric acid without iso-acids. Production of lactic acid varies within the genus *Fusobacterium* but *Leptrotrichia* produces a major lactic acid peak. Some of the species, including *F. necrophorum,* can also convert threonine or lactate to propionate, characteristics that are useful for identification. The genus *Bacteroides* has recently undergone redefinition and consolidation, and currently includes organisms that produce acetic and succinic acids, with small amounts of isobutyric and isovaleric sometimes produced. Two exceptions are *B. splanchnicus* and *B. putredinis,* which also produce butyric acid. These organisms likely will be renamed in the near future. One of the new genera created for organisms previously found in *Bacteroides* is *Prevotella*. *Prevotella* species have the same end products as *Bacteroides*. Another new genus is *Porphyromonas,* whose members produce butyric acid in addition to iso-acids, but not succinic acid. Production of phenylacetic acid distinguishes *P. gingivalis* from *P. asaccharolytica*. End products of other gram-negative organisms less frequently encountered in clinical specimens are listed in Table 5-1.

The genus *Peptostreptococcus* (which now includes all clinically significant species from the former genus *Peptococcus* except for *Peptococcus niger*) is not defined by specific end products; however, the species within the genus do have characteristic patterns. Thus, the morphologically and biochemically similar *P. magnus* and *P. prevotii* may be differentiated by production of butyric acid by *P. prevotii* and only acetic acid by *P. magnus. P. anaerobius* is easily identified by the presence of isocaproic acid in culture media.

The three genera of gram-negative cocci have unique end-products. *Veillonella* produces acetic and propionic acids, *Acidaminococcus* forms acetic and butyric acids, and *Megasphaera* produces iso-acids plus butyric, valeric and caproic acids.

Although an organism's metabolic pathway is a genetically stable characteristic, various factors can influence the end products measured. An important variable is the ratio of the concentration of carbohydrate to peptone in the medium in which the organism is grown.[242-244] Glucose is metabolized preferentially over peptones by organisms capable of fermentation. Thus, in a peptone-glucose broth, end products of glucose metabolism, such as acetate and butyrate, will accumulate first. Iso-acids, formed by the metabolism of peptones, will appear more slowly. If an organism is a non-fermenter, the iso-acids will be present in young cultures. Media rich in peptones, such as chopped meat broth, will yield larger amounts of iso-acids than media poor in peptones, such as thioglycolate broth. Quantities of end products increase with the age of the culture until the nutrients are depleted and growth ceases, so quantities of any product within the same species can vary accordingly. Chopped meat broth contains large amounts of lactate (from the meat); thus any organism capable of converting lactate to propionate (such as *F. necrophorum*) will yield a large propionic acid peak when grown in chopped meat broth that is absent when the organism is grown in peptone-yeast-glucose or thioglycolate broths. The presence of CO_2 will enhance production of succinate by some species of *Actinomyces*.

Pankuch and coworkers [188] have utilized a method for determining end-products directly from pure cultures on agar plates. Two 3-5 mm diameter plugs are removed from an area of good growth on agar into two separate glass tubes, one for volatile acids and one for non-volatile acids. One-tenth of a milliliter of 50% (v/v) H_2SO_4 is added to each tube. The tubes are heated gently over a flame to melt the agar. The liquid is then treated as a broth culture and end products may be extracted and measured.

Formic acid is produced by many organisms, but we have not found its presence or absence to be of any help in identification of clinical isolates. Since it is not detected in aqueous samples, we usually disregard it. It can be detected in ether extracts of samples analyzed on a chromatograph equipped with a thermal conductivity detector.

Culture media may contain trace amounts of some of these acids. Therefore, one should assay an uninoculated incubated broth and subtract any end products detected from those found in the samples.

CHROMATOGRAPHS, COLUMNS, AND OPERATING CONDITIONS

The separation of fatty acids by GLC is achieved by injecting the prepared sample through a septum into the proximal end of a long column packed with special material contained in a heated oven. The high temperature in the injector port volatilizes the sample and the flow of a carrier gas moves the sample down the length of the column. The fatty acids are retained for different time periods within the column by the packing material, according to their molecular weight and polarity. When a fatty acid reaches the detector at the distal end of the column, it causes an electrical response which is converted by the recorder to a peak on the chart. The more fatty acid present, the larger the peak. The retention time (the length of time from the injection of the sample to the detection of the peak) determines the identity of the peak, as compared to a standard containing defined quantities of the same fatty acids as the end products of anaerobic metabolism. Increasing the carrier gas flow rate decreases the retention time for all peaks as they are forced out more rapidly; however, peak separation (resolution) also decreases. Increasing the oven temperature has the same effect.

Figure 7-1 depicts the patterns of the volatile and non-volatile fatty acid standard mixtures used as controls for GLC analysis of end products of anaerobic metabolism. Control standards are available from Supelco and other sources.

The most basic and least expensive type of gas chromatograph is equipped with a thermal conductivity detector (TCD). It requires only helium as a carrier gas and can be used to measure both volatile and non-volatile end products. Care must be exercised to exclude water from the sample because it increases the size of the injection peak and may overlap a lactic acid peak. Flame ionization detector (FID) chromatographs do not detect water; thus aqueous culture supernatants can be analyzed directly. However, they require three different gases for operation: nitrogen as the carrier gas and air and hydrogen for the flame. There may be restrictions in some institutions on storage of pure hydrogen in the laboratory. Alternatively, a hydrogen-generating instrument produces hydrogen from water by electrolysis in sufficient amounts for use in flame ionization detectors. This compact instrument is made by Packard and is available through Supelco.

Gas chromatographs suitable for analysis of metabolic end products are available in a wide range of sophistication and price. They may include an automatic injector or an

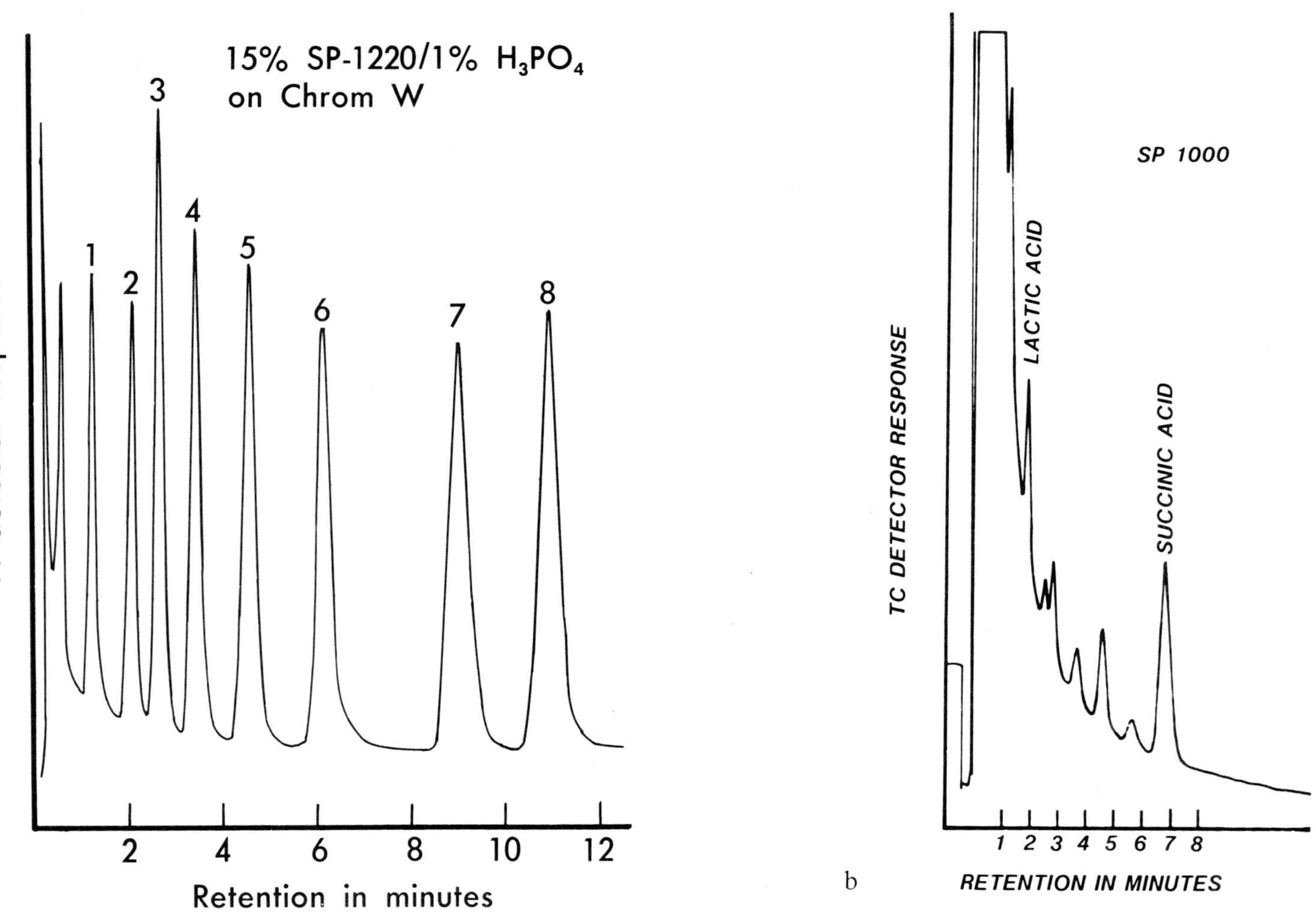

Figure 7-1.   GLC traces obtained with operating parameters listed in table 7-1. a. volatile fatty acid mixture in aqueous solution (FID). Peaks: 1, acetic acid; 2, propionic acid; 3, isobutyric acid; 4, n-butyric acid; 5, isovaleric acid; 6, n-valeric acid; 7, isocaproic acid; 8, n-caproic acid; b. methyl esters of non-volatile acids in chloroform (TCD).

integrator and computer which convert peak areas to milliequivalent (mEq) quantities. The price will vary accordingly. Highly sophisticated equipment is not required for end product analysis where only a semi-quantitative result is needed.

Table 7-1 summarizes columns with which we have had experience. When injecting aqueous samples into the SP-1220 column, note that ghost peaks may appear from the previous sample, especially with older columns. This can be partially remedied by occasionally injecting methanol to "wash" the column. Also, some organisms, especially clostridia, produce alcohols that may elute near fatty acid peaks. If this occurs, the chromatogram obtained from the test organism should be carefully compared to that of the standard fatty acid mixture to accurately identify the peaks. Alternatively, an additional standard that contains these alcohols can be utilized to identify these peaks.

Prepacked columns are available from several sources. We use the Supelco SP-1000 and SP-1220. Both columns may be used to separate the volatile short chain, one to seven carbon molecules, fatty acids and the methyl esters of the non-volatile fatty acids. The SP-1220 is the faster of the two columns, allowing for a shorter analysis time. SP-1000 elutes alcohols well in advance of the fatty acids so that there is less likelihood of overlapping peaks. SP-1220 gives poorer separation of the alcohols from acetic acid with samples extracted into ether.

The operating parameters listed in Table 7-2 serve as a guide. We have found that a new column operates efficiently with a slightly lower carrier gas flow rate and a lower oven temperature. As a column is used, residues may clog the entry and efficiency may be improved by increasing the carrier gas flow rate and the injector, oven, and detector temperatures, always being careful not to exceed the maximum temperature tolerated by the column packing, as specified by the manufacturer. When changes in operating parameters are made, it is essential to analyze the standard mixture to verify that all peaks are present and well separated.

PREPARATION OF SAMPLES FOR ANALYSIS

Incubate the test organism in a broth medium until good growth is achieved. Optimum quantities of end products are usually produced after 48 hours with anaerobes that grow well. Acidify the broth culture to pH 2.0 by adding 2 drops of 50% H_2SO_4 to 5 ml of the

Table 7-1. Suggested columns

Column	Detector	Volatile fatty acids detected	Non-volatile fatty acids detected	Other
SP-1000[1]	TCD	Yes	Yes	Formic
SP-1220[2]	FID	Yes	Yes	

[1] 10% SP-1000/1% H_3PO_4 on chrom W A W, 100-120 mesh, Supelco

[2] 15% SP-1220/1% H_3PO_4 on chrom W A W, 100-120 mesh, Supelco

Column packing	FID SP 1220	TCD SP 1000
Gases	N₂ 30 ml/min H₂ 25 ml/min. Air 300 ml/min	He 100 ml/min
Injection temp. (°C)	150 - 180	150 - 180
Detector temp. (°C)	150 - 180	150 - 180
Oven temp. (°C)	140 - 170	140 - 170
TCD filament	—	200 m amp.
attenuation	—	1:2 - 1:4
FID range	10	—
attenuation	x 128 - 8	—
Recorder range	1 mv full scale	1 mv full scale
Chart speed	1/2"/min or 1 cm/min	1/2"/min or 1 cm/min

broth. Centrifuge briefly to pellet the cell mass and remove the supernatant for analysis. Once acidified, the end-products remain stable and the supernatant may be stored in the refrigerator for several months.

Volatile fatty acids

The volatile fatty acids produced by anaerobic bacteria include formic, acetic, propionic, n-butyric, isobutyric, n-valeric, isovaleric, n-caproic, isocaproic, and heptanoic.

FID. For detection in a chromatograph equipped with an FID and an SP-1220 column, inject 1 μl of the supernatant without further processing.

TCD. **Note: work in a fume hood.** If a chromatograph equipped with a TCD is used, extract the sample as follows:

1. Pipet 1 ml of supernatant into a 13 x 100 mm glass tube.
2. Add 0.2 ml of 50% H_2SO_4, 0.4 g NaCl, and 1 ml methyl tert-butyl ether (methyl tert-butyl ether is preferable to ethyl-ether because it has a higher boiling point and is easier to handle).
3. Stopper the tube with a butyl rubber stopper and mix well by inverting approximately 20 times. Remove the stopper after 10 inversions to release excess pressure. Do not use Parafilm because it will solubilize in the ether.
4. Centrifuge briefly or allow to stand for a few minutes to separate the water/ether layers.
5. Remove the ether layer (top) to another tube and add a pinch of 4-20 mesh anhydrous $CaCl_2$ to remove all traces of water. Shake the tube gently and allow to stand for a few minutes. The sample is now ready for injection.

6. The amount of sample to be injected varies with the column and operating conditions. One should inject varying amounts of the standard mixture (extracted in the same fashion) to determine the optimum amount. It may vary from 5 to 15 μl. The injection peak should not obscure any of the peaks to be detected.

Non-volatile fatty acids

The non-volatile organic fatty acids useful for identification include lactic, succinic and phenylacetic acids. For detection, their methyl derivatives must first be prepared as follows:

1. Pipet 1 ml of supernatant into a 13 x 100 mm glass tube.
2. Add 1 ml methanol and 0.1 ml 50% H_2SO_4.
3. Stopper tube, shake or vortex vigorously and heat to 60°C for 30 minutes, or allow it to stand at room temperature overnight.
4. Add 0.5 ml chloroform, replace stopper and mix well. If the sample was heated, allow to cool to room temperature before adding chloroform. **Note: work in a fume hood.**
5. Allow to stand for a few minutes, or centrifuge briefly to separate the chloroform/water layers.
6. Draw the bottom chloroform layer into the syringe, being careful not to take up any water, wipe the outside of the needle, and inject into the column. The amount of sample to be injected will vary with the column and operating conditions. Too much sample may obscure significant peaks.
    Note: Always test the standard solution, methylated in the same manner, to determine the optimum amount to inject. The range is usually 1 to 14 μl.

OTHER APPLICATIONS OF END PRODUCT ANALYSIS

End product analysis may also be performed directly on pus or infected body fluids. For optimum results, ether extracts should be prepared from these samples. This procedure can sometimes provide rapid presumptive identification of anaerobes present, or can serve to distinguish an anaerobic from an aerobic infection. If many types of organisms are present, it may be difficult to determine which organism is producing which end product. For example, if gram-positive cocci and gram-negative bacilli are found on the Gram stain, and butyric acid is detected, one can infer that *Peptostreptococcus prevotii, P. asaccharolyticus, P. tetradius,* and/or *Fusobacterium* sp. are present. If butyric acid is not detected, one could infer that other *Peptostreptococcus* sp. or facultative gram-positive cocci are present, and that the gram-negative bacillus is not a *Fusobacterium.* If gram-negative and gram-positive bacilli are noted on the Gram stain, and iso-acids are detected, they may be from a *Bacteroides* sp., or a *Clostridium* or *Eubacterium* sp. It has been our experience that a careful reading of the Gram stain usually yields as much information as end product analysis when several types of organisms are present. If only one organism is present, such as in blood culture bottles, end product analysis can provide rapid presumptive identification of an anaerobic organism.[266] End product analysis can also be a useful aid for diagnosis of bacterial vaginosis.[220]

WHOLE CELL LONG CHAIN FATTY ACID METHYL ESTER (FAME) ANALYSIS

A recently introduced method developed for identification of bacteria is based on analysis of whole cell fatty acid methyl ester profiles.[30,86] These fatty acids are typically 9-20 carbons in length and usually reside in the cell membrane, unlike the short chain fatty acids formed as a result of metabolism. The most common system used is the Hewlett Packard Microbial Identification System (HP 5898A MIS). It consists of a 5890A gas chromatograph containing a 25m x 0.2 mm methyl-phenyl-silicone fused silica capillary column, an automatic sampler with sample tray, an integrator, a mini-computer and a printer. The software is available from MIDI (Microbial ID Inc.). A fairly extensive data base for anaerobes has recently been compiled by the VPI Anaerobe Laboratory.

The organism to be analyzed is grown in a defined medium for a specified length of time (usually $\leq$ 48 h). The cells are harvested, then saponified. Fatty acids are thus cleaved from the cell lipids and are converted to their sodium salts. The fatty acids are then methylated to form fatty acid methyl esters, which are more volatile, for GC analysis. The fatty acid methyl esters are extracted into an organic solvent phase, which is washed to remove residues that could damage the chromatographic system before being introduced into the instrument. The peaks obtained from the sample are compared to profiles of well-characterized strains in the data base and the best match identification with a statistical likelihood is given. The precision of identification varies from species to species.

The specific reagents and methods are fully described in a manual provided by the manufacturer.

Susceptibility Testing of Anaerobic Bacteria

The National Committee for Clinical Laboratory Standards (NCCLS) has recently published a revised approved standard for susceptibility testing of anaerobic bacteria.[183] The methods include agar dilution testing, microbroth dilution, and macrobroth dilution. Although broth disk elution was approved in the past, the committee members felt that the unreliability of the method and the poor correlation of the results with the reference method did not warrant its inclusion as an approved alternative method.

GENERAL CONSIDERATIONS

Anaerobic susceptibility testing need not be performed routinely on all isolates in the clinical laboratory. Table 8-1 suggests the types of infections for which anaerobic susceptibility testing should be considered and other indications for susceptibility testing, and Table 8-2 lists the isolates that should be considered for testing. Otherwise, clinicians and clinical microbiologists may be guided by published reports of the efficacy of many new agents, but should be aware that inter-laboratory variations in technique, as well as the inherent variability in the methods and local strain differences may lead to differing susceptibility patterns. Methodological differences may sometimes result in extreme differences in test results, as with ceftizoxime, which shows excellent activity against anaerobes by broth microdilution tests, and poorer activity with agar dilution tests. For a more detailed review of anaerobic susceptibility testing, see Wexler.[252]

Susceptibility testing of anaerobes may be undertaken in a clinical setting with individual isolates, or as a large study to evaluate antimicrobial agents and report current resistance patterns. In either case, the following caveats should be kept in mind:

- Unless adequate growth of the organism is achieved, the MIC cannot be reliably determined. Many anaerobes grow poorly in microdilution tests utilizing Wilkins-Chalgren broth and in agar dilution tests using Wilkins-Chalgren agar.
- MICs for control organisms should fall within the acceptable range with each antimicrobial agent. MICs of quality control organisms in a broth microdilution technique are often one two-fold dilution lower than in an agar dilution technique.

Studies to determine QC MICs for broth microdilution are being carried out.

- Modifications of the methods (e.g., higher inoculum, longer incubation time) and media (e.g., added supplements) may permit the growth of most fastidious organisms, but such modifications should be undertaken only when necessary and with appropriate quality control. Reports should state what modifications of the reference method were used to obtain the result, especially if the data are to be published.
- Results will vary depending on the method used. Care should be taken when comparing results generated by use of different methods. As stated above, broth microdilution methods tend to produce MICs slightly lower than MICs obtained by agar dilution methods.
- The MICs of certain antimicrobial agents cluster near the breakpoints for those agents. Since the acceptable error for the dilution methods described (except for the spiral gradient method) is one two-fold dilution, many strains may be designated as resistant on one occasion and susceptible on another, **within the allowable error of the technique.**
- Determination of endpoint is probably one of the greatest sources of error in interpretation of susceptibility tests (Figure 8-1). The problem exists to some extent with most gram-negative anaerobic bacteria. It may be particularly troublesome with those organisms that grow poorly, so that the growth control itself is fairly light. In some instances, there may be evidence of light growth (or multiple tiny colonies) persisting in the presence of very high concentrations of drug, despite a sharp drop-off in growth at a much lower drug level. In the case of *Fusobacterium,* this has been shown to be due to the persistence of cell-wall deficient forms of the bacteria.[123] Recent experiments have indicated that the haze seen with *B. wadsworthia* is also due to spheroplast formation, and indeed, this may be the cause of the haze seen with many gram-negative organisms and beta-lactam agents, all of which affect bacterial cell wall formation. In these cases, the point at which the growth drops off sharply should be read as the MIC, and the persistence of haze noted. In the case of some organisms (e.g., *B. gracilis,* some pigmented *Prevotella* or *Porphyromonas,* some anaerobic cocci, and *B. wadsworthia*), there may be no sharp drop-off of growth, and

Table 8-1. Indications for susceptibility testing for anaerobes

---

A. Specific infections from which isolates should be considered for susceptibility testing

    Refractory or recurrent bacteremia

    Central nervous system infections

    Endocarditis

    Osteomyelitis

    Joint infection

    Prosthetic device infection

    Organism isolated from any normally sterile site

    Infection not responsive to empiric therapy

B. To determine patterns of susceptibility in a particular hospitals or geographic area; these should be done at intervals of four to six months.

C. Evaluation of new antimicrobial agents

Table 8-2. Isolates to consider for testing

Bacteroides fragilis group isolates	*Clostridium perfringens*
B. gracilis *	*C. ramosum, C. innocuum, C. clostridioforme*
Fusobacterium mortiferum/varium	*Bilophila wadsworthia* *

* Supplements recommended

the endpoint may be very difficult to determine. Reading the plate against a background of transmitted light (rather than reflected light) may be helpful. Determination of endpoint in some cases is necessarily arbitrary, and undue importance should not be attributed to these results. We are currently investigating the use of a viablity indicator dye, tetrazolium chloride (TTC), to aid in endpoint determination.

When susceptibility testing is performed for the purpose of evaluation of agents or determining antibiograms, a few additional cautions are necessary:

- Recent clinical isolates should be used to determine current antibiograms. Patterns of antimicrobial susceptibility may change in an institution over time.
- Adequate numbers of strains (at least 10 of each species) should be tested if inferences are to be made regarding the susceptibility of a particular group or species.
- Appropriate proportions of clinically relevant species should be used. *B. fragilis,* for example, is much more susceptible than are the other members of the group, yet the other species account for at least half of the clinical isolates. Ideally, the different species of the *B. fragilis* group should be reported separately.

PROCEDURES FOR TESTING SUSCEPTIBILITY OF ANAEROBIC BACTERIA

Agents to be tested

The hospital formulary should serve as a guide for choosing the agents for testing. If susceptibility testing is not being performed routinely for anaerobes, it should be undertaken when a patient is non-responsive to empiric therapy or when there is a reason to suspect resistance to one of these agents; the laboratory should include the agent being

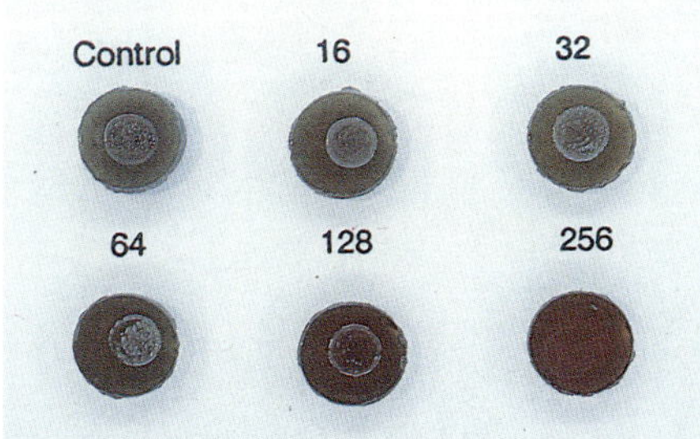

B. distasonis on ceftizoxime
agar dilution plates.

Figure 8-1. Illustration of an endpoint which is difficult to determine. *B. distasonis* on agar dilution plates containing concentrations of ceftizoxime (μg/ml) as indicated. Note that photography exaggerates the appearance of the "haze" seen at the higher concentrations.

Table 8-3. Susceptibility of Anaerobes to Antimicrobial Agents*

Percent susceptible	*B. fragilis* Species	*B. fragilis* Group (excluding *B. fragilis*)	*B. gracilis* †	Other *Bacteroides*	*Fusobacterium*
>95% susceptible	Ampicillin+Sulbactam Piperacillin+Tazobactam Ticarcillin+Clavulanate Cefoperazone+Sulbactam Imipenem Chloramphenicol Metronidazole	Ampicillin+Sulbactam Piperacillin+Tazobactam Ticarcillin+Clavulanate Cefoperazone+Sulbactam Imipenem Chloramphenicol Metronidazole	Imipenem Chloramphenicol	Piperacillin Ampicillin+Sulbactam Ticarcillin+Clavulanate Cefoperazone Cefotaxime Cefoxitin Cefoperazone+Sulbactam Ceftizoxime Imipenem Chloramphenicol Clindamycin	Penicillin G Piperacillin Ampicillin+Sulbactam Piperacillin+Tazobactam Ceftizoxime Imipenem Chloramphenicol Clindamycin Metronidazole
85-95% susceptible	Cefotetan Cefoxitin Clindamycin Piperacillin	Piperacillin	Metronidazole	Cefotetan Ceftazidime Ceftriaxone	Ticarcillin+Clavulanate Cefoperazone Cefotaxime Cefotetan Cefoxitin Ceftriaxone Cefoperazone+Sulbactam
70-84% susceptible	Ceftizoxime Moxalactam	Cefoxitin Clindamycin	Piperacillin Ampicillin+Sulbactam Moxalactam Cefoperazone+Sulbactam		Ceftazidime Moxalactam
50-69% susceptible	Cefoperazone Cefotaxime Ceftazidime Ceftriaxone	Cefoperazone Cefotetan Ceftizoxime Moxalactam	Penicillin G Ticarcillin+Clavulanate Cefoperazone Cefotaxime Cefotetan Cefoxitin Ceftizoxime Clindamycin		
<50% susceptible	Penicillin G	Penicillin G Cefotaxime Ceftazidime Ceftriaxone	Ceftazidime	Penicillin G Moxalactam	

PERCENT OF STRAINS SUSCEPTIBLE AT BREAKPOINT

>95% susceptible   85-95% susceptible   70-84% susceptible   50-69% susceptible   <50% susceptible

The ranking within a color group does not reflect degree of activity; drugs are arranged by classes.

* These data represent a compilation of studies from the Wadsworth Anaerobe Laboratory.

† Strains which were previously identified as *B. gracilis* will probably be split into several different taxa. Also, if one uses formate/fumarate (a growth supplement for *B. gracilis*) in the test medium, many strains will have considerably lower MICs with certain antibiotics. The susceptibility patterns will also vary among the new taxa.

(Modified from Finegold, Baron and Wexler, *A Clinical Guide to Anaerobic Infections,* 1992. Courtesy Princeton Scientific Productions, Inc. and Star Publishing Co.)

Susceptibility of Anaerobes to Antimicrobial Agents*

Bilophila	Peptostreptococcus	Clostridium perfringens	Other Clostridium species	Nonspore-forming Gram-positive rods
>95% susceptible Chloramphenicol Metronidazole Imipenem Ticarcillin Ticarcillin+Clavulanate Cefoxitin	**>95% susceptible** Penicillin Piperacillin Ampicillin+Sulbactam Ticarcillin+Clavulanate Cefoperazone Cefotetan Ceftazidime Ceftriaxone Moxalactam Cefoperazone+Sulbactam Imipenem Chloramphenicol	**>95% susceptible** All drugs are active at ≥95% level	**>95% susceptible** Amoxicillin Ampicillin Carbenicillin Penicillin G Piperacillin Ticarcillin Ampicillin+Sulbactam Imipenem Chloramphenicol Metronidazole	**>95% susceptible** Penicillin G Piperacillin Ampicillin+Sulbactam Ticarcillin+Clavulanate Cefotaxime Ceftizoxime Imipenem Chloramphenicol
85-95% susceptible Clindamycin	**85-95% susceptible** Metronidazole		**70-84% susceptible** Cefoxitin Moxalactam Clindamycin	**85-95% susceptible** Cefotetan Cefoxitin Ceftazidime Ceftriaxone Cefoperazone+Sulbactam Clindamycin
70-84% susceptible Cefotaxime Cefotetan	**70-84% susceptible** Clindamycin		**50-69% susceptible** Cefoperazone Cefotaxime Ceftizoxime Ceftriaxone	**70-84% susceptible** Cefoperazone Moxalactam
<50% susceptible Penicillin G Ceftizoxime			**<50% susceptible** Ceftazidime	**50-69% susceptible** Metronidazole

PERCENT OF STRAINS SUSCEPTIBLE AT BREAKPOINT

>95% susceptible	85-95% susceptible	70-84% susceptible	50-69% susceptible	<50% susceptible

The ranking within a color group does not reflect degree of activity; drugs are arranged by classes.

* These data represent a compilation of studies from the Wadsworth Anaerobe Laboratory.

used therapeutically. Imipenem, chloramphenicol, metronidazole and the β-lactam/β-lactamase inhibitor combinations are presently almost uniformly active against anaerobes in the U.S. Of course, these agents must be monitored periodically in large centers in order to detect any developing resistance. Many cephalosporins, penicillins and clindamycin are variably active against anaerobes, and one cannot predict the susceptibility of a particular strain based on published patterns. Norfloxacin, ofloxacin and ciprofloxacin are relatively poor in activity against anaerobes; temafloxacin had better activity[253] but is no longer available. Some of the newer quinolones being developed have a much improved in *vitro* spectrum against most anaerobes.

Current patterns of susceptibility of anaerobic bacteria tested at the Wadsworth Anaerobe Laboratory are summarized in Table 8-3. In an effort to simplify the presentation and interpretation, antibiotics with similar efficacies against groups of anaerobes have been grouped and the range of percent susceptible strains has been used to define the group. The ranking within a group does not reflect the degree of activity. This presentation also minimizes the significance of differences of $< 10\%$. At times an antibiotic was borderline between two groups, and the decision about its grouping was necessarily arbitrary.

Preparation of inoculum

A McFarland turbidity standard should be used for preparation of inocula for susceptibility tests. Instructions for making McFarland standards are found in Appendix C. The 0.5 McFarland standard should correlate with approximately 1.5×10^8 CFU/ml, but will vary widely for organisms of different sizes. Scientific Devices Inc. manufactures McFarland standards using latex beads (also distributed by Remel); these may be stored and used indefinitely. We have had no experience with these standards.

There are two acceptable methods for preparing the inoculum for susceptibility testing.

(1) Direct suspension: Use a sterile cotton swab to suspend colonies from a 24-72 hour blood agar plate directly into a clear broth (such as Brucella) to achieve turbidity equal to a 0.5 McFarland standard.
(2) Growth suspension:

    a. Inoculate 5 or more colonies into enriched fluid thioglycolate medium (or other appropriate broth) and incubate 4-6 hours (or overnight for slow-growing organisms).
    b. Dilute to the density of a 0.5 McFarland standard with a clear broth.

Preparation of antimicrobial stock solutions and dilutions

Stock solutions of antimicrobial agents generally may be prepared in advance and frozen in aliquots at -70°C. After thawing they should not be refrozen. Some antibiotics (imipenem, for example) are not very stable upon freezing (even at -70°C) and should be made fresh. The solvents and diluents suitable for many antibiotics are indicated in Table 8-4. Contamination is very rare, and solutions generally need not be sterilized, although sterile water and buffers should be used for making the solutions. If desired, solutions may be sterilized through a nitrocellulose membrane filter. Other filters such as paper, asbestos, or sintered glass may adsorb appreciable amounts of some agents and should not be used. Generally, unless the manufacturer indicates otherwise, cephalosporins and cephamycins may be dissolved in 0.1 M phosphate buffer (PB), pH 6.0, and further diluted in water.

Either of the following formulae may be used to determine the amount of powder or diluent needed for a standard solution:

$$\text{Weight (mg)} = \frac{\text{Volume (ml) x Concentration needed } (\mu g/ml)}{\text{Assay potency } (\mu g/mg)}$$

or

$$\text{Volume (ml)} = \frac{\text{Weight (mg) x Assay potency } (\mu g/mg)}{\text{Concentration needed } (\mu g/ml)}$$

Table 8-4. Solvents and diluents for antimicrobial agents

Antimicrobial Agent	Solvent[1]	Diluent
Amoxicillin, Ticarcillin	0.1 M PB[2], pH 6	0.1 M PB, pH 6
Ampicillin, Cefoperazone	0.1 M PB, pH 8	0.1 M PB, pH 6
Azlocillin, Carbenicillin	Water	Water
Cefamandole, Cefmetazole, Cefotaxime	Water	Water
Cefotetan	DMSO[1] or DMF[1]	Water
Cefoxitin, Ceftizoxime	Water	Water
Cephalothin	0.1 M PB, pH 6	Water
Chloramphenicol	95% ethanol[3]	Water
Clavulanic acid	0.1 M PB, pH 6	0.1 M PB, pH 6
Clindamycin	Water	Water
Erythromycin	95% ethanol[3]	Water
Imipenem[4]	0.01 M PB, pH 7.2	0.01 M PB, pH 7.2
Metronidazole	DMSO	Water
Mezlocillin	Water	Water
Moxalactam (diammonium salt)[5]	0.04 N HCl	0.1 M PB, pH 6
Penicillin G, Piperacillin	Water	Water
Sulbactam	0.1 M PB, pH 8	0.1 M PB, pH 6
Tetracycline	Water	Water
Vancomycin	0.05 N HCl[6]	Water

[1] Dissolve powder in minimum amount of solvent needed to solubilize agent and bring to volume with diluent. DMSO = dimethyl sulfoxide, DMF = dimethyl formamide

[2] Phosphate buffer

[3] Agent is dissolved in 1/10 final volume 95% ethanol, then brought to the proper volume with water.

[4] Imipenem is not very stable and should be freshly prepared.

[5] Let stand for 1.5 to 2 hours to allow the solution to come to an equilibrium of the two types of isomers that are present in solutions for clinical use.

[6] Stock solution is made with 0.05 N HCl and further dilutions are made with water.

Prepare the dilutions according to the scheme in Table 8-5. Avoid serial two-fold dilutions since any error made toward the beginning of the dilution scheme will be compounded. The dilution scheme shown in Table 8-5 will minimize the chance of error carryover.

Addition of growth supplements to antimicrobial susceptibility test media

Unless good growth of the organism is achieved, an accurate MIC cannot be determined. "Growth stimulation test" (Appendix B) and Box 5-1 list some possible supplements for specific groups of organisms; in each case, the microbiologist must ascertain that the supplement does not interfere with the test by ensuring that the quality control MICs fall within acceptable values. We recommend that formate/fumarate routinely be added to the media (agar or broth) for testing of *B. gracilis* and other *B. ureolyticus* group strains; the light, nearly transparent growth of the organism in conventional media resulted in our reading erroneously high MIC's for certain strains. Other supplements may be added, such as 0.1% Tween 80 for gram-positive cocci. Rabbit or horse serum or 2-3% laked

Table 8-5. Preparations of dilutions (for agar dilution tests)

To:	Add:	Concentration (μg/ml)	Concentration in agar plate (μg/ml)[1]
Stock		2560	256
2 ml 2560	2 ml water	1280	128
1 ml 2560	3 ml water	640	64
1 ml 2560	7 ml water	320	32
2 ml 320	2 ml water	160	16
1 ml 320	3 ml water	80	8
1 ml 320	7 ml water	40	4
2 ml 40	2 ml water	20	2
1 ml 40	3 ml water	10	1
1 ml 40	7 ml water	5	0.5
2 ml 5	2 ml water	2.5	0.25
1 ml 5	3 ml water	1.25	0.125
1 ml 5	7 ml water	0.625	0.0625
2 ml 0.625	2 ml water	0.31	0.031
1 ml 0.625	3 ml water	0.15	0.015

[1] When 2 ml added to 18 ml agar

horse blood may improve the growth of pigmented *Prevotella, Porphyromonas* and other fastidious anaerobes. Pyruvate (1%) is recommended for testing *Bilophila* sp. Again, since supplements may interfere with testing, it is imperative that the quality control organisms remain within their accepted range.

Quality control standards

The use of quality control strains is necessary to monitor the quality of the media and the stability of the antimicrobial preparations. Appropriate control tests should be performed whenever new media, reagents, or antimicrobials are prepared. **Include at least two quality control strains with each test run.**

The NCCLS approved quality control strains and the acceptable MIC ranges (obtained by the agar dilution technique) for these strains are listed in Table 8-6. MIC values for both macro- and microbroth dilution are often one two-fold dilution step lower, and values in agar media containing blood may be one two-fold dilution step higher than the NCCLS-approved values. Studies are underway to determine appropriate quality control values for microdilution tests.

Agar dilution test procedure

In advance (up to one month):

1) Prepare the medium (i.e., the Wilkins-Chalgren or Brucella base) in the amount needed for a test run and autoclave; keep refrigerated, and **do not store longer than one month.** For the supplemented Brucella base, add the vitamin K_1 and hemin prior to autoclaving.

Up to 24 hours preceding the test:

1) If media is prepared in advance, melt and cool the medium to 48°C in a water bath. If laked blood is to be added, it may be added at this point.
2) Prepare dilutions of antimicrobial agents (Table 8-5).
   Note: The new NCCLS-Approved Standard includes the use of Brucella laked blood agar as an approved technique; we have found that this medium gives the best growth of many fastidious organisms, including *Bacteroides gracilis,* many pigmenting *Prevotella* and *Porphyromonas, Fusobacterium,* anaerobic cocci, and *Bilophila wadsworthia.*
3) Incorporate dilutions into either Wilkins-Chalgren agar (for the NCCLS reference technique) or into supplemented Brucella laked blood agar. Add 2.0 ml of each antimicrobial dilution to a tube containing 18 ml of molten agar ($\sim$48°C) to make a 1:10 dilution. Mix by gently inverting the tube twice and pour into a petri dish.
4) After plates have solidified, place them in a 35-37° incubator with the top of the plates slightly ajar for 30 to 45 min. to allow evaporation of excess moisture. Plates may be refrigerated if the test is to be performed the next day (except plates containing imipenem).

Performing the test run:

1) Remove the plates from the refrigerator and allow to warm to room temperature.
2) Add the individual inocula (diluted to a 0.5 McFarland density) to the wells of a replicator device (such as a Steers replicator).

Table 8-6. Acceptable Ranges of MIC's for Control Strains Using NCCLS Reference Method

Antimicrobial Agents	*Bacteroides fragilis*[1] ATCC 25285	*Bacteroides thetaiota-omicron*[1] ATCC 29741	*Clostridium perfringens*[1] ATCC 13124	*Eubacterium lentum*[1,2] ATCC 43055	*Peptostreptococcus magnus*[3] ATCC 29328
Amoxicillin/clavulanate	0.25-1	0.5-2			
Ampicillin	16-64	16-64	NR[5]		0.125-0.25
Ampicillin/sulbactam	0.5-2	0.5-2			
Carbenicillin	16-64	16-64	0.25-1		0.25-1
Cefamandole	32-128	32-128	0.06-0.25		
Cefmetazole[4]	8-32	32-128	NR	4-16	
Cefoperazone	32-128	32-128	NR	32-128	
Cefotaxime	8-32	16-64	0.06-0.25	64-256	
Cefotetan	4-16	32-128	NR	32-128	
Cefoxitin	4-16	8-32	0.25-1	4-16	0.125-0.5
Ceftizoxime	32-128	NR	NR	16-64	
Ceftriaxone[4]	32-128	64-256	NR		
Chloramphenicol	2-8	4-16	2-8		2-4
Clindamycin	0.5-2	2-8	0.03-0.12	0.06-0.25	0.5-1
Imipenem	0.03-0.12	0.06-0.25	0.03-0.12	0.25-1	
Metronidazole	0.25-1	0.5-2	0.12-0.5		0.25-2
Mezlocillin	16-64	8-32	0.06-0.25	8-32	
Moxalactam	0.25-1	4-16[7]	0.03-0.12[7]	64-256	
Penicillin G[6]	16-64 8-32	16-64 8-32	0.06-0.25 (0.03-0.12)		0.063-0.125
Piperacillin	2-8	8-32	0.06-0.25	8-32	
Tetracycline	0.12-0.5	8-32	0.03-0.12		0.25-1
Ticarcillin	16-64	16-64	0.25-1		
Ticarcillin/clavulanate	NR	0.5-2	0.12-0.5		

[1] NCCLS-approved control strains

[2] Data for *E. lentum* is considered tentative for a one year period following publication of NCCLS M11-A2.

[3] *Peptostreptococcus magnus* is not an NCCLS-approved quality control strain. These figures were obtained from data at the Wadsworth Anaerobe Laboratory using the Wadsworth method and may be useful for laboratories needing quality control at low concentrations.

[4] Data considered tentative for one year after publication of NCCLS document M11-A2.

[5] NR--no MIC range recommended

[6] Penicillin values are in units/ml (μg/ml in parentheses).

[7] Values valid for agar reference method only; different results may be obtained with other methods, media, or enrichments. Results with media containing blood tend to be one two-fold step higher and results from broth dilution tests one two-fold step lower than above. Aside from entries with the "7" superscript, values apply to either Wilkins-Chalgren or Brucella blood agar.

3) Stamp the inocula on the antimicrobial-containing plates, beginning with the lowest concentration of antimicrobial agent and proceeding to the highest. Try to stamp the least effective antimicrobial agents first.
   Note: If several agents known to have potent anti-anaerobe activity are to be tested, it is best to stamp from different inoculum blocks to avoid any antibiotic carryover effect.
4) At the beginning and end of each set of antibiotic plates, stamp two plates with no antibiotic; incubate one anaerobically (growth control) and one aerobically (aerobic contaminant control.) An additional plate should be inoculated and subsequently refrigerated to be used as an inoculum control (to distinguish slight growth from dried inoculum.)
5) Leave the plates at room temperature until the inoculum is absorbed (approximately 10 minutes).
6) Stack the test plates and incubate upside down (to prevent condensate from falling on the inoculated spots) in an anaerobic atmosphere at 35-37° for 48 hours.
7) Determine the MIC (see below).

Reading endpoints. The MIC is defined as the lowest concentration of drug yielding no growth, a haze, one discrete colony or multiple tiny colonies (Figure 8-2). In the case of persistent light growth, the MIC is read at the concentration at which there is a *marked* change as compared to the growth control. See "General Considerations," above, for a detailed discussion of problems related to endpoint determination. In large research evaluations of antimicrobial agents, difficult endpoints readings should be especially noted, since they may cause considerable discrepancies in results. Breakpoints approved by the National Committee for Clinical Laboratory Standards (NCCLS) are listed in Table 8-7.

The use of triphenyltetrazolium chloride.(TTC) as an aid in determination of endpoint. With some organism/antibiotic combinations, endpoints "trail" and are difficult to read. In recent studies, we investigated the use of tetrazolium chloride (a dye which is reduced by many viable bacteria to a red compound, formazan) as an aid in endpoint determination. We found that endpoints determined with the aid of TTC

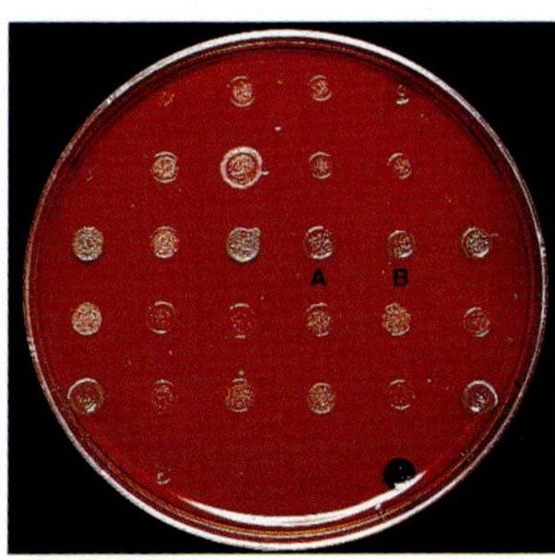

Plate 1. Growth Control.

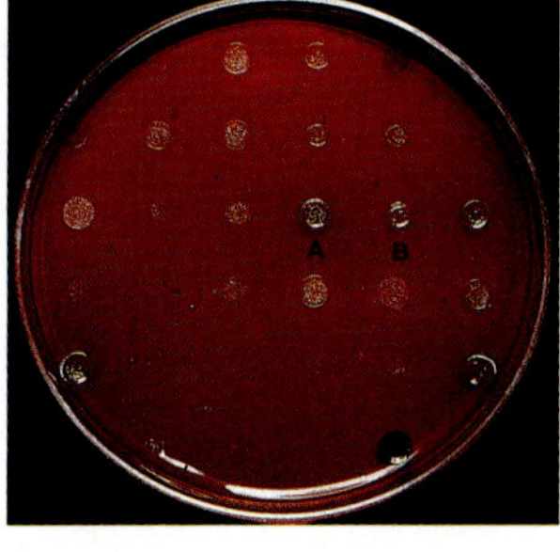

Plate 2. Antimicrobial Agent (16 μg/ml).

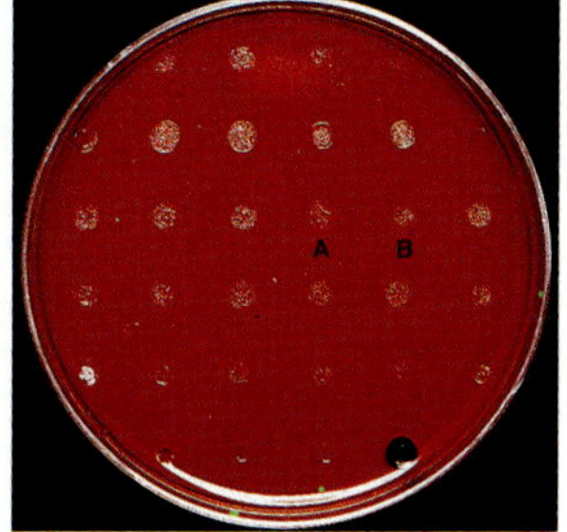

Plate 3. Antimicrobial Agent (32 μg/ml).

Figure 8-2. Agar dilution susceptibility testing. (From Finegold, Baron, and Wexler, *A Clinical Guide to Anaerobic Infections,* 1992. Courtesy Princeton Scientific Productions, Inc. and Star Publishing Co.)

Table 8-7. NCCLS-Approved Breakpoints for Antimicrobial Agents

Antimicrobial Agent	Breakpoint MIC (μg/ml)
Amoxicillin/clavulanate	8/4
Ampicillin	4[1]
Ampicillin/sulbactam	16/8
Carbenicillin	128
Cefamandole	16
Cefmetazole	32
Cefoperazone	32
Cefotaxime	32
Cefotetan	32
Cefoxitin	32
Ceftizoxime	64[2]
Ceftriaxone	32
Chloramphenicol	16
Clindamycin	4
Imipenem	8
Metronidazole	16
Mezlocillin	64
Moxalactam	32[3]
Penicillin G	4[1]
Piperacillin	64
Tetracycline	8
Ticarcillin	64
Ticarcillin/clavulanate	64/2

[1] This breakpoint was chosen because it divides the population of β-lactamase producing strains from the population of non-β-lactamase producers. However, the actual blood levels achievable are higher, so that infection with non-β-lactamase producers with higher MICs might respond to therapy. Use nitrocefin test for β-lactamase activity for isolates with MICs$\geq$1 μg/ml.

[2] For agar dilution only. For broth microdilution, the breakpoint is 32 μg/ml.

[3] Breakpoint may be lower with doses of $\leq$4 g/day.

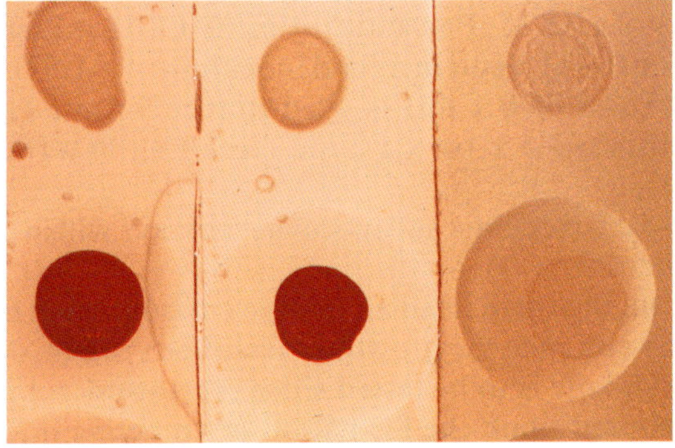

Figure 8-3. Growth of *Bilophila wadsworthia* on agar dilution plates containing imipenem. Top row: *B. wadsworthia* on agar dilution plates. Note that determination of an endpoint would be very difficult. Bottom row: *B. wadsworthia* on agar plates overlayed with triphenyltetrazolium chloride solution. Viable organisms produce a red color. Note how endpoint determination is facilitated.

correlated with the "viability endpoints" measured by determining the concentration of antibiotic which permitted no net growth of organisms on Steer's replicator "spots" over a 48 hour period. We are continuing to investigate the use of viability indicator dyes in aiding MIC determinations (Figure 8-3).

Broth microdilution tests

A number of manual, semiautomated, and automated devices are available which allow rapid and economical preparation of large numbers of plates containing two-fold dilution series of antimicrobial agents. Plastic (microtiter) trays containing eight by twelve rows of small flat-bottomed, U-or V-shaped wells are used.

In advance:

1) Prepare antimicrobial agent dilutions in Schaedler's, Wilkins-West, Brucella, brain heart infusion or Anaerobe broth (Difco Laboratories; a broth made with the same formulation as Wilkins-Chalgren agar but without the agar.) Prepare the antimicrobial dilutions in at least 10 ml of broth.

2) Dispense 0.1 ml of the antimicrobial solutions into the microtiter plates using either a manual or automated dispensing device.
   Note: The trays should not contain less than 0.1 ml per well. Volumes of less than 0.1 ml are not recommended for anaerobes because of evaporation, which can result in concentration of antimicrobial agents, and because of the marked effect of the higher inoculum size in smaller volumes.

3) Seal the prepared trays in plastic bags and freeze at -70°C until needed. Such trays usually remain stable for 4-6 months. **Freezers with self-defrosting units should not be used.** Appropriate quality control should be performed to monitor shelf life.

Alternatively, frozen or lyophilized plates containing diluted antimicrobial agents to which inoculum is added may be purchased (see below). At least one well should contain broth with no drug to serve as a growth control.

Performing the test run:

1) Remove the trays from the freezer and allow them to thaw.
2) Prepare the inoculum as described earlier. The final inoculum in each well should be 1×10^5 CFU. **Note: The volume of the inoculum is generally 0.01 ml (10 μl); an**

inoculum adjusted to the turbidity of a 0.5 McFarland ($\sim$1.5x10^8 CFU/ml) and diluted 1:10 (i.e., $\sim$1.5x10^7 CFU/ml) will result in a final inoculum of $\sim$1.5x10^5 CFU/well. If diluted inoculum (usually 0.1 ml) is used to reconstitute lyophilized trays, the broth must contain 1.5 x 10^6 CFU/ml (resulting in a final inoculum of $\sim$1.5x10^5 CFU/well).

3) Inoculate trays. Prepare dilutions of the inoculum and inoculate the trays within 15 minutes after the inoculum is standardized.

Note: Reducing the plates (i.e., placing them in an anaerobic environment) for 2-4 hours prior to inoculation may enhance the growth of fastidious organism. Plates must be reduced if metronidazole is to be tested since the antimicrobial activity of metronidazole is dependent upon the formation of an active intermediate which demands a reduced atmosphere. Manufacturer's guidelines should be followed. Different commercial suppliers may give different guidelines as to whether plates should be inoculated in the chamber or on the bench.

4) Stack trays in multiples of four or less, to allow even incubation temperature. The top tray may be covered with a plastic lid to prevent evaporation. Cellophane tape should *not* be used unless the plates were inoculated inside an anaerobic chamber.

5) Incubate the trays in an anaerobic atmosphere for 48 hours.

6) Read the MIC as the lowest concentration of antimicrobial agent that completely inhibits the growth of the organism (Figure 8-4).

Note: In the case of trailing endpoints (gradually diminishing growth), the concentration that produces the most significant reduction of growth as compared to the growth control well should be chosen as the endpoint. With metronidazole, a tiny button of precipitate may occur in all wells. The endpoint again is read as the concentration in the well where a drastic change in the size of the button occurs. Quality control strains should be included with each test run; the values tend to be one two-fold step lower than those listed for the reference agar dilution method.

There are some commercial sources for frozen or lyophilized trays. Micro-Media Systems, Microtech Medical Systems, and Innovative Diagnostic Systems sell frozen trays for use with anaerobes. Radiometer/Sensititre produces lyophilized trays that may be adequate for use with anaerobes if suitable medium is added. The limited number of appropriate antimicrobials active against anaerobes in some of these trays remains a drawback. Microtech Medical Systems, PML Microbiologicals and Sensititre will supply laboratories with custom-poured MIC panels. When using commercial trays,

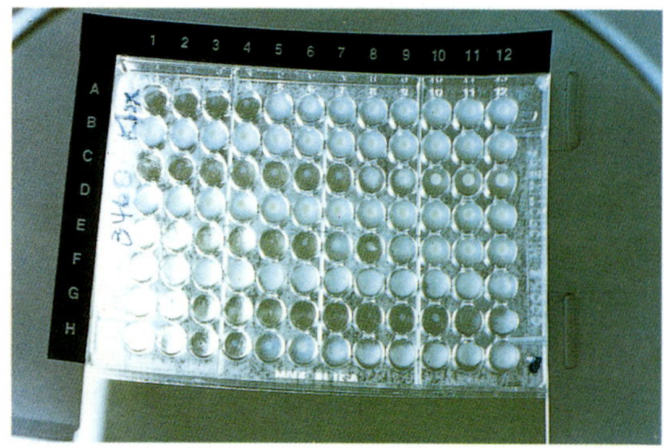

Figure 8-4. Microdilution plate. The MIC is defined as the concentration in the first well showing no button of growth.

follow the manufacturer's directions for test performance and quality control, while ensuring that the basic principles mentioned above are maintained (adequate inoculum and final volume).

Although some workers have attempted to determine minimum bactericidal concentrations (MBCs) using microbroth trays, this cannot be done reliably and we do not recommend it.

Broth macrodilution tests

Conventional broth dilution tests are useful for testing organisms with swarming growth (such as some *Clostridium* spp.) or if MBCs must be determined. (Testing of swarming organisms may also be accomplished with the microdilution technique.) Serial two-fold dilutions of antimicrobial stock solutions are prepared in 2.5 ml of Brucella broth containing hemin (5 μg/ml), $NaHCO_3$ (1 mg/ml) and vitamin K_1 (1 μg/ml), or one of the broths mentioned in the microdilution procedure. Tubes should be prepared within 3 hours of use, or else frozen at -70°C.

The inoculum should be a 1:200 dilution of a 0.5 McFarland standard (prepared as described above and made in the same broth used to dilute the drugs). A final inoculum volume of 2.5 ml is added to the broth containing the drug (the final inoculum will be $\sim$3 x 10^5 CFU/ml). Incubate the tubes in an anaerobic atmosphere at 37°C for 48 hours. Include an inoculated broth containing no antimicrobial agent as a growth control for each strain tested. Include a tube of uninoculated broth with each day's tests to serve as a negative growth control. Also include tests on control strains, results of which should fall within appropriate reference values. Values for broth tests will often be one dilution lower than for agar dilution tests. The MIC is read as the lowest concentration showing no visible growth.

MBCs (minimum bactericidal concentrations) may be determined by streaking 0.1 ml from each tube (after allowing 48 hours for growth and recording MICs) to a blood agar plate. Plates are incubated anaerobically for 48 hours and the MBC is read as the lowest concentration of drug resulting in fewer than 30 colonies (99.9% killing rate).

Other methods of susceptibility testing

Spiral gradient endpoint system. The spiral streaker (Spiral Systems Instruments) deposits a measured amount of antimicrobial stock solution in a spiral pattern on an agar plate, resulting in a radially decreasing concentration gradient (from the center of the plate). After allowing the antimicrobials to diffuse for 3-4 hours, the isolates (prepared to a 0.5 McFarland standard) are deposited on the plate using an automated inoculator, or manually streaked from the edge to the center of the plate. The plates are incubated for 48 hours in an anaerobic atmosphere, after which the endpoints of growth are marked and the distance in mm measured from the center of the plate to the point where growth stops or where there is a distinct reduction in growth (Figures 8-5a and 8-5b). The data are then entered into a computer using a software program provided by the manufacturer that determines the concentration of drug at that point based on the radius of growth and the molecular weight (i.e., diffusion characteristics) of the antimicrobial agent. Details of the procedure may be found in the manufacturer's guidelines. Some workers have compared

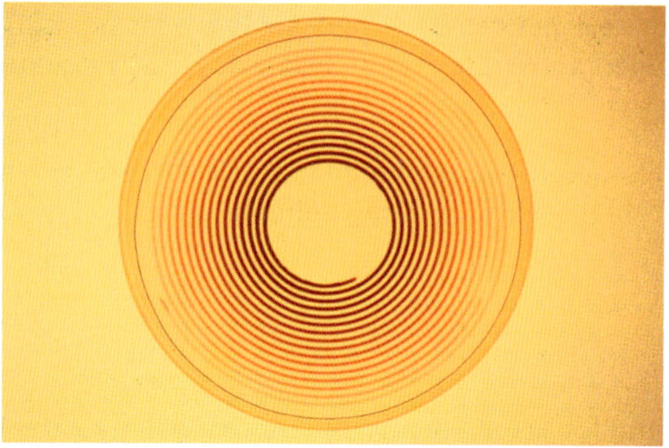

Figure 8-5a. A spiral gradient on the surface of a petri plate formed with crystal violet dye to illustrate the antibiotic gradient formed by a spiral plater (photograph courtesy of Spiral Systems Instruments, Bethesda, MD).

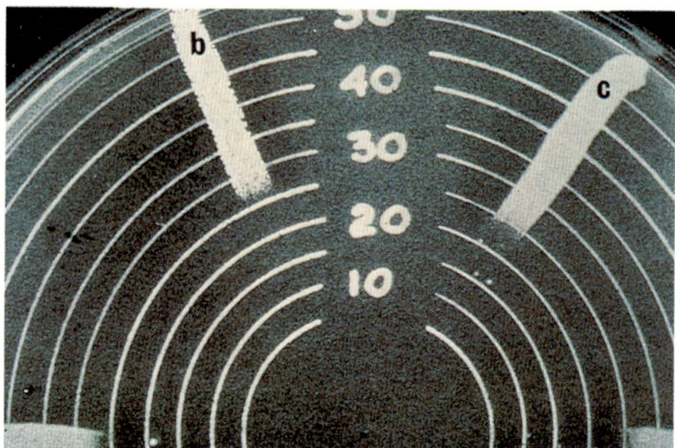

Figure 8-5b. Organisms inoculated in radial lines on an agar plate containing an antibiotic gradient created by the spiral plater. The plate is viewed through the Spiral Gradient Endpoint (SGE) template for measurement of the endpoint location. Spiral gradient endpoints are measured as the distance along the radius from the center of the plate to the point of growth inhibition (photograph courtesy of Spiral Systems Instruments).

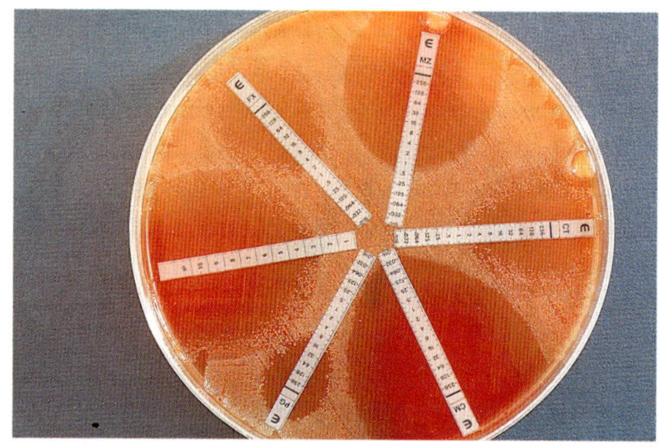

Figure 8-6. Example of the E test for determining the susceptibility of an anaerobic bacillus to six antimicrobial agents. The MIC is read from the point of the filter paper at which the teardrop-shaped zone of inhibition ends.

this procedure with standard agar dilution [107] and found good correlation, as we have in our own tests.[254]   The technique retains the advantages of an agar dilution system (particularly good growth of fastidious organisms) as well as some of the labor-saving advantages of the simpler techniques. The spiral gradient system shows approximately 20% variability, which is much lower than the variability seen with agar dilution susceptibility tests. This method is especially suited for batch testing and would not be particularly efficient for testing single isolates against a variety of antimicrobials.

E test.   This test produced by AB Biodisk uses a plastic strip coated with an antibiotic gradient on one side and an MIC interpretive scale on the other. The strip is applied to the surface of an agar plate which has been inoculated with a 0.5 McFarland suspension of the organism, and the plate is incubated anaerobically for 24 to 48 hours depending on the growth rate of the organism. The point at which the teardrop-shaped zone of inhibition intersects the interpretive scale is read as the MIC (Figure 8-6). The Wadsworth Anaerobe Laboratory has not evaluated these strips, although one of us has studied their use for anaerobic susceptibility testing with 105 strains with 98% categorical agreement with the agar dilution reference method.[49]   This test is particularly suited for studying individual patient isolates against a few selected agents; testing large numbers of isolates or a wide range of antimicrobials would not be accomplished efficiently with this method.

Agar disk diffusion tests.   Barry *et al.*[13] have evaluated modifications of the agar disk diffusion technique for use with rapidly growing anaerobes. While agar diffusion tests correlated well with the standard technique for fast-growing anaerobes for some anti-microbial agents, there were other cases in which the correlation was poor. At present, the test is not considered appropriate for anaerobic susceptibility testing.

β-lactamase testing.   The β-lactamase test is simple and rapid and may be used as a supplement to conventional susceptibility tests. It may provide the microbiologist with some useful information. If a strain is β-lactamase positive, it will probably be resistant to the β-lactamase labile agents (e.g., many penicillins and cephalosporins) and susceptible to the β-lactam agents when combined with β-lactamase inhibitors such as sulbactam, tazobactam, or clavulanic acid. However, if a strain does not produce β-lactamase, one cannot assume that it will be susceptible to the β-lactam agents. *Bacteroides gracilis* and many strains of *Bacteroides distasonis* do not produce β-lactamase, yet are quite resist-ant to β-lactam agents. β-lactamase-producing anaerobes are listed in Table 8-8.

The most reliable method for detecting β-lactamases (specifically cephalosporinases) in anaerobes utilizes nitrocefin disks (Cefinase; BBL). The disk should be moistened with sterile water, and then several well-isolated colonies should be smeared on the disk with a sterile loop or stick. A positive reaction is indicated by a color change from yellow to red. Most reactions occur within 5 to 10 minutes but there are some β-lactamase-positive strains that react more slowly (up to 30 minutes).

Table 8-8. β-lactamase producing anaerobes[1]

Bacteroides fragilis group	*Porphyromonas* sp.
B. coagulans	*Fusobacterium nucleatum*
B. splanchnicus	*F. mortiferum/varium*
Pigmented *Prevotella*	*Megamonas hypermegas*
P. oralis	*Mitsuokella multiacida*
P. bivia	*Clostridium ramosum*
P. disiens	*C. clostridioforme*
P. buccae	*C. butyricum*
P. oris	*Bilophila wadsworthia*

[1] Some strains of these species produce β-lactamases

Appendix A

Special Procedures for Specimen Collection and Research Studies

---

LABORATORY HANDLING OF ORAL SITE SPECIMENS TO AVOID INDIGENOUS ORAL FLORA

General Principles

The saliva and normal mucosal surfaces of the oral cavity contain anaerobic bacteria, streptococci, and other indigenous commensal flora in very large numbers. Certain of these bacteria contribute to pathologic processes when they are able to colonize a different environmental niche aided by local disease, trauma, poor oral hygiene, or other infectious processes. The clinical significance of any bacteria recovered from specific oral sites can be assessed only when contamination by the indigenous oral flora has been avoided.

Numerous strategies for collection of site-specific specimens from within the oral cavity have been developed.[84,89,151,174,175] Transcutaneous aspiration through iodine-disinfected, uninvolved skin is still the most desirable collection method for obtaining specimens from oral infections pointing toward the skin surface. Several collection methods are presented in this section along with a general scheme for processing oral site specimens for quantitative anaerobic bacteriology.

Organisms in gingival sites are present as microcolonies, often held together by complex polysaccharide matrices (glucans, etc.); these aggregates must be dispersed to obtain reliable quantitative results. Sonication or physical disruption using a Vortex mixer and glass beads are usually employed. For transport of all oral specimens, Hungate-stoppered tubes or vials containing prereduced, sterile anaerobic diluent (e.g., one-quarter strength Ringer's solution with 0.1% sodium metaphosphate) in 0.5-2.0 ml volumes with 6-8 glass beads (1.0-1.5 mm in diameter), may be used. Another transport system, VMGA III,[140,177] utilizes sterile semisolid anaerobic transport medium (2 ml/screw-cap tube) filled almost to the brim of the transport tube (Figure A-1).

Figure A-1. VMGA III transport vial for dental specimens. Note glass beads and medium filled almost to the top of the vial.

Saliva specimen collection

1. Instruct the patient to expectorate saliva into the sterile tube, either after paraffin stimulation or without stimulation. A minimum of 2.0 ml is usually collected.
2. Immediately draw the saliva into a syringe.
3. Inject the saliva sample into the anaerobic transport tube.
4. Deliver to the laboratory for immediate processing.

Supragingival plaque specimen collection

1. Isolate the area to be sampled with cotton rolls and dry the sample site with sterile cotton swabs.
2. Collect a sample of supragingival or coronal plaque with the curette or scaler.
3. Open the cap of the transport tube and immediately insert and swirl the scaler or curette tip into the transport medium. Do not open transport tube until specimen is ready to be inserted. Minimize exposure to air! Close the tube immediately. If a gassing cannula is available, maintain continuous flow of oxygen-free gas into the tube.
4. Transport the tube to the laboratory for processing as soon as possible.

Subgingival plaque specimen collection

1. Isolate the sample site with cotton rolls and dry the area with sterile cotton swabs.
2. Remove supragingival plaque with sterile toothpicks or with a sterile scaler.

If using the sterile scaler:
3. Obtain subgingival plaque with scaler or curette.
4. Immediately transfer and swirl the scaler tip into the anaerobic transport vial. If a gassing cannula is available, maintain continuous flow of oxygen-free gas into the tube when the top is open.
5. Tighten the cap and deliver to the laboratory for processing.

If using paper point method:
3. Retract the gingival margin and insert three paper points to the depth of the sulcus and hold for 10 seconds.
4. Without delay, transfer the paper points into the anaerobic transport vial. If a gassing cannula is available, maintain continuous flow of oxygen-free gas into the tube when the top is open.
5. Tighten the cap and deliver to the laboratory for processing.

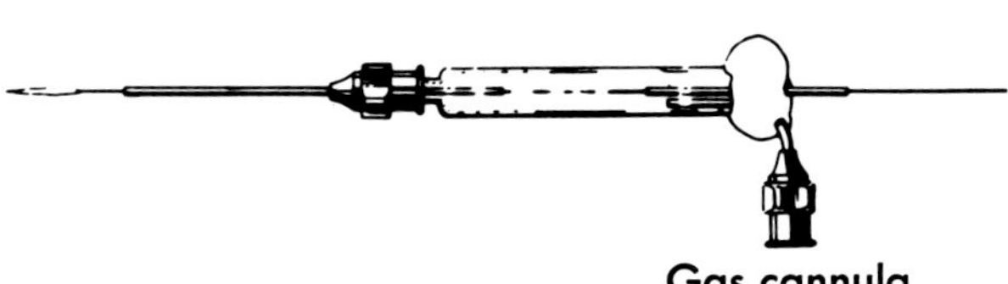

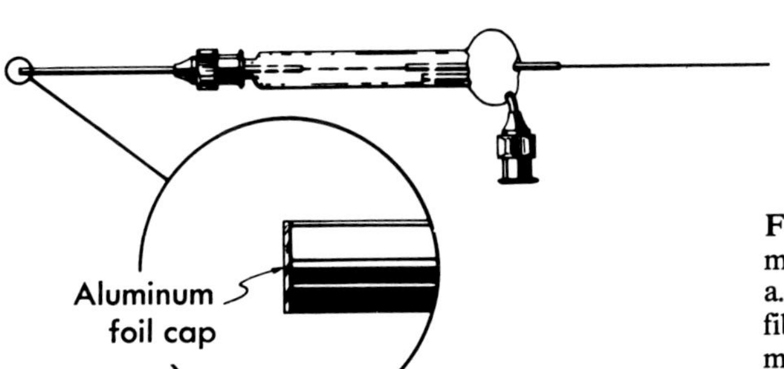

Figure A-2. Sampling devices for gingival specimens.

a. The barbed broach is wound with calcium alginate fibers and inserted into a syringe with the plunger end modified to permit continuous flushing with oxygen-free gas. The tip of the needle is capped with aluminum foil and sterilized.

b. In use, the needle is placed into the sampling site, and the barbed broach is passed through the needle, dislodging the aluminum cap. The sample is taken and the broach drawn back into the needle, and the device removed from the site. The aluminum is removed by means of a scaler. (From Newman, M.G., and others: J. Periodont. 47:373-379, 1976. By permission of American Academy of Periodontology).

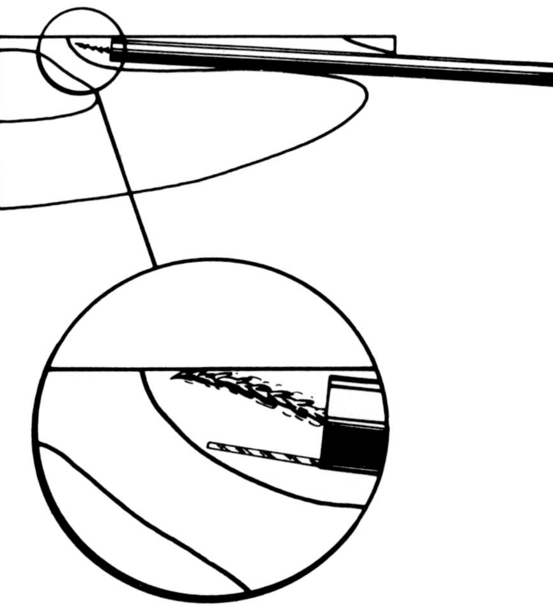

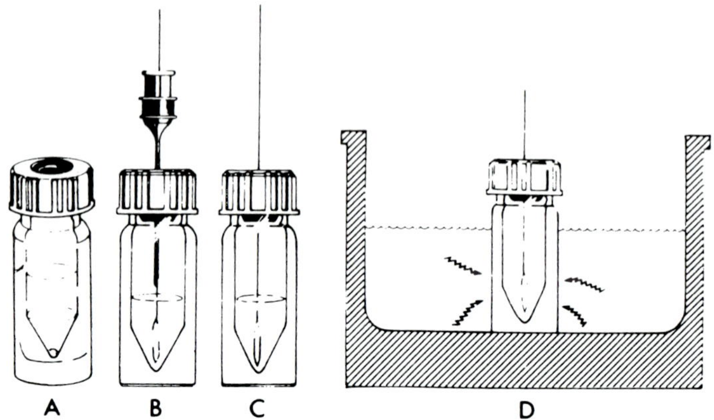

Figure A-3. Specimen handling for subgingival plaque specimens collected on barbed broach. a. Screw capped vial with diaphragm. B. Broach containing sample inserted through needle through the diaphragm. c. Needle removed from diaphragm leaving broach inside. d. Sonicating sample prior to dilution.

If using the barbed broach:

3. Retract the gingival margin and insert the needle to the depth of the sulcus.
4. Pass the barbed broach through the needle, dislodging the aluminum foil cap.
5. Obtain specimen (Figure A-2) and retract the broach back into the needle.
6. Without delay, insert the needle containing the broach through the septum of the transport vial and extend the broach into the medium (Figure A-3).
7. Remove the needle, leaving the broach inside the vial.
8. Deliver to the laboratory for processing.

Processing oral specimens in the laboratory

Specimen processing

1. Perform all manipulations of the specimen in an anaerobic chamber if possible. Quantitative studies of oral flora cannot be performed adequately in room air.
2. If the specimen is to be sonicated (transport medium without beads), place the entire transport vial into the sonicator for 30 seconds before manipulating the sample in the chamber.
3. If the transport medium contains glass beads for dispersal of the specimen, warm the vial briefly to liquefy the agar and mix the vial or tube vigorously on a Vortex mixer for a minimum of 20 seconds (30 seconds is recommended) before further manipulations in the chamber.
4. Make serial ten-fold dilutions in PRAS Ringer's solution or other appropriate broth to 10^{-6}. These dilutions should allow enumeration of all isolates on at least one set of plates. Metaphosphate (0.1%) may be added to the first dilution tube to prevent reaggregation of the organisms.
5. Spread 0.1 ml of inoculum from each dilution onto the surface of each agar medium being used for the study. Table A-1 contains suggested media. Inoculum can be spread using a pipette to run the suspension slowly onto the surface as the plate is rotated mechanically, or manually using a sterile wire or glass spreader, or with the aid of the Spiral Streaker instrument.

   Note: For darkfield microscopy, an additional sample is taken and suspended in sterile saline with 1% gelatin. The sample should be examined while still warm; the bacteria are divided into groups according to morphology and motility.

Table A-1. Suggested media for oral flora studies [1]

Medium	Incubation atmosphere	Dilutions plated[2]	Days of incubation	Purpose
Brucella PRAS blood agar [3]	Anaerobic	10^{-3}, 10^{-4}, 10^{-5}, 10^{-6}	4-7	Total counts and predominant flora
TSBV agar	Air + 5-7% CO_2 & Anaerobic	Undiluted, 10^{-1}	2-7	*Actinobacillus actinomycetemcomitans*
KVLB agar [4]	Anaerobic	Undiluted, 10^{-1}, 10^{-2}	4-7	*Bacteroides*, Pigmented *Prevotella*
JVN or NV agar	Anaerobic	Undiluted, 10^{-1}, 10^{-2}	4-7	*Fusobacterium* and *Leptotrichia*
Lactobacillus selective medium	Anaerobic	10^{-1}, 10^{-2}	2-3	*Lactobacillus*
BGSA agar	Anaerobic	Undiluted, 10^{-1}	4-7	*B. gracilis*
Veillonella neomycin agar	Anaerobic	10^{-1}, 10^{-2}	2-7	*Veillonella* and other gram-negative cocci
CFAT agar	Air + 5-7% CO_2	Undiluted, 10^{-1}	4-7	*Actinomyces*
Chocolate-bacitracin agar	Air + 5-7% CO_2	10^{-1}, 10^{-2}, 10^{-4}	1-2	*Haemophilus*, *Campylobacter*, and other gram-negative capnophiles
Mitis-Salivarius agar	Air + 5-7% CO_2	10^{-2}, 10^{-4}, 10^{-6}	1-2	Alpha- and nonhemolytic streptococci
Brucella or trypticase soy blood agar	Air + 5-7% CO_2	10^{-2}, 10^{-4}, 10^{-6}	1-2	Total counts and predominant capnophilic flora

[1] See Appendix A for interpretive information on media.

[2] 0.1 ml of each dilution is placed on each medium used.

[3] Anaerobic media should ideally contain 2 different nonselective basal media, e.g., Brucella + FAA, or Brucella + TSA.

[4] Modification of KVLB containing 2 μg/ml vancomycin (KVLB-2) may be helpful for selection and isolation of *Porphyromonas* sp.

Incubation and examination. Table A-1 lists suggested incubation conditions and periods for different media used. It also lists the organisms isolated on different selective media. Descriptions of typical colonies on selective agars are found at the end of this Appendix ("Descriptions of Typical Colonies on Selective Media for Anaerobes"). For identification of anaerobic organisms refer to Chapters 4 and 5; the facultative and aerobic organisms may be identified as described in the Manual of Clinical Microbiology [9] and Diagnostic Microbiology. [10]

RESPIRATORY SECRETION SPECIMENS VIA THE PROTECTED BRONCHIAL BRUSH CATHETER TO AVOID ORAL CONTAMINATION (SUITABLE FOR ANAEROBIC CULTURE)

General Principles

Anaerobic bacteria are often involved in infections of the lung and pleural space, particularly in patients who have aspirated oral or gastric secretions or who have significant periodontal disease. [76] Because normal oral secretions contain large numbers of the same organisms that may become pathogens when they reach the normally sterile alveoli, specimens for **definitive** bacteriologic evaluation must be collected so as to bypass oral flora. Only two methods are well established currently: percutaneous transtracheal aspiration and the use of protected bronchial brush in a double-lumen catheter. [34,37,76] Transtracheal aspiration is seldom done presently. Performed correctly, the protected bronchial brush catheter (PBC) culture collection is well tolerated and should take less than 2 minutes. [37] Bronchoalveolar lavage (BAL) has not yet been studied adequately but appears promising. Both PBC and BAL specimens must be cultured quantitatively.

Respiratory specimens collected via the upper airway and thus contaminated with saliva and other oral secretions (such as sputum and endotracheal suction specimens) may be evaluated for the presence of non-anaerobic organisms known to be pathogens: *Staphylococcus aureus, Streptococcus pneumoniae, Haemophilus influenzae, Pseudomonas, Moraxella (Branhamella) catarrhalis,* and Enterobacteriaceae, **all of which may simply be colonizing the respiratory tract and not be the etiologic agent of pneumonia.** Thus, a culture report indicating the recovery of any of these organisms does not establish the diagnosis. Respiratory cultures should be evaluated in relationship to the clinical illness of the patient.

Certain etiologic agents, such as viruses, are not easily detected in respiratory secretions. For diagnosis of disease due to these agents, the protected bronchial brush catheter (PBC) is recommended. [43]

Patients who might be considered **poor** candidates for PBC cultures are those who are unable to cooperate with medical personnel, who have significant hypoxia, or who have unstable cardiac conditions or untreated asthma.

Handling the specimen after collection

Note: proper handling is essential to achieving meaningful results.

1. Vigorously wipe the outer surface of the inner cannula (which is extending from the outer cannula) with 70% alcohol.

2. Using sterile scissors, cut off the distal end of the inner cannula (right up to the tip of the brush). This cannula tip should be discarded.
3. Then extend the brush through the cut end of the cannula and aseptically cut it off so that it falls into a tube containing 1.0 ml of transport medium. If only one sterile scissors is available, it should be flamed in the alcohol burner before it is used to cut off the brush. The cap of the tube should be tightened securely.
4. Hold the tube of transport medium containing the brush at room temperature and deliver to the laboratory immediately.

Laboratory processing of the PBC specimen

Media

1. 2 tubes of thioglycolate broth, 9.9 ml each
2. Agar plates for subculture, including 3 BAP, 3 KVLB, 3 PEA, or other anaerobic plates used. The aerobic plates, Choc, BAP, and MacConkey, are optionally placed in the chamber or on the bench.

Specimen processing:

1. Initial processing should be performed in the anaerobic chamber. If the specimen is to be sonicated (transport medium without beads), place the entire transport vial into the sonicator for 30 seconds before manipulating the sample in the chamber.
2. If the transport medium contains glass beads for dispersal of the specimen, warm the vial briefly to liquefy the agar and mix the vial or tube vigorously on a Vortex mixer for a minimum of 20 seconds (30 seconds is recommended) before further manipulations.
3. Use a 1.0 ml pipette and draw up enough of the specimen suspension to place 0.1 ml on each agar plate being inoculated, 0.1 ml into the first of the two 9.9 ml thio tubes (to make a 1:100 dilution of the original specimen), and a large drop on each necessary slide. Label the plates "10^{-1}." Label the thioglycolate "10^{-2}."
4. Slides should always be prepared for Gram stain.
5. Spread the inoculum on the plates using a pipette to run the suspension slowly onto the surface as the plate is rotated mechanically, or manually using a sterile wire or glass spreader, or with the aid of the Spiral Streaker instrument.
6. Mix the 1:100 dilution thioglycolate tube thoroughly (invert or vortex) and use a fresh, sterile 1.0 ml pipette to repeat the procedure carried out on the original specimen. These plates, prepared from 0.1 ml of the 1:100 dilution should be labeled "10^{-3}." The second 9.9 ml thioglycolate tube should be labeled "10^{-4}."
7. Streak the second set of plates as before.
8. Thoroughly mix the second thioglycolate tube, and use a fresh, sterile pipette to place 0.1 ml on each of one final set of agar plates. Label these plates "10^{-5}." This is the final dilution.
9. Incubate the anaerobic plates in the chamber. Pull all other plates, the thioglycolate broths, and slides out of the chamber for appropriate handling. Alternatively, all non-anaerobic media and slides may be prepared on the bench from the thioglycolate tubes after the anaerobic plates have been inoculated in the chamber.

Interpretation of results

1. Hold anaerobic plates for 7 days before discarding as negative. Hold other plates as for routine respiratory specimens. Examine thioglycolate broth and subculture only if

there is no growth on any agar plates or if Gram stain suggests an organism not recovered on plates.

2. The number of CFU in the original specimen is determined by counting colonies and multiplying by the dilution factor. (Example: If 24 colonies of staphylococci are present on the plates labeled "10^{-3}," then the final CFU will be 2.4 x 10^4/ml original specimen.) Since the PBC is expected to hold only 0.01 ml, this amount of growth corresponds to 2.4 x 10^6 CFU/ml in the patient's secretions. For another style of PBC that holds only 0.001 ml, the 24 colonies would correspond to 2.4 x 10^7 CFU/ml. CFU counts greater than 1,000/ml in the specimen (and thus $>10^5$/ml in the secretions) are considered to be significant. Count only the plates containing countable colonies (30-300). Do complete ID and susceptibilities on all organisms present in $>$1,000/ml in the original specimen.

3. All colony morphotypes in less than significant numbers should be identified morphologically only.

COLLECTION OF ENDOMETRIAL SPECIMENS USING THE PIPELLE PROTECTED SUCTION CURETTE

General Principles

The best diagnostic specimen for determining the microbial etiology of endometrial infections is a sterile tissue specimen obtained with minimal or no contamination with vaginal flora and avoiding atmospheric oxygen (to allow recovery of anaerobic bacteria).[64] If the specimen is contaminated from endogenous sources, the culture results may be uninterpretable. Because the infecting agents are usually found within tissue, a routinely collected swab specimen is almost never adequate for obtaining relevant microbiological data.[164]

Specimen collection (See Figure A-4)

1. Collect the specimen as described by the manufacturer.
2. Remove the screw-cap on the transport tube, aseptically cut off the distal tip of the Pipelle sheath, just behind the hole, and expel the collected material into the vial. **To depress the plunger of the piston, you must feed it into the sheath by holding it close to the sheath opening and pushing it in 1 inch at a time.**
3. Immediately replace the screw-cap and close tight.
4. Maintain at room temperature. Transport to the laboratory within 2 hours of collection.

Specimen processing

1. Process the specimen in the anaerobic chamber if possible. If a chamber is not available, work quickly on the bench top within a safety cabinet, as aerosols are possible.
2. Use a wide-tipped sterile pipette to transfer the endometrial tissue into a sterile tissue grinder. Add thioglycolate broth if necessary to add volume. Alternative method: use a Stomacher (Tekmar Co.) to homogenize the material.
3. Homogenize or grind the specimen to a smooth consistency. Use a pipette to place a drop or two on the surface of each agar plate and into thioglycolate enrichment broth.

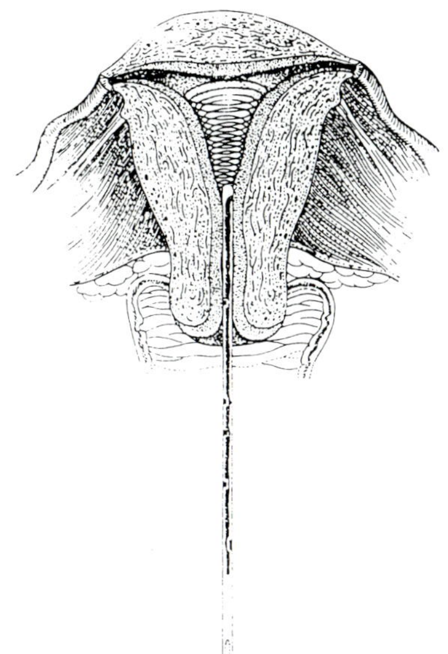

Figure A-4. Cutaway diagrammatic view through the uterus during collection of endometrial contents with a Pipelle endometrial suction device. The vacuum created by drawing back on the plunger draws tissue into small openings at the end of the curette as the device is twirled circularly and moved laterally. (Drawing by Charles Stern, courtesy of Unimar, Wilton, CT.)

Suggested media include: anaerobic blood agar, PEA, BBE, and KVLB agar plates for anaerobic isolation and 5% sheep blood, MacConkey, Thayer-Martin or Martin-Lewis, chocolate, and Columbia with colistin and nalidixic acid (CNA) agar plates for aerobic isolation. *Mycoplasma* medium (Shepard's A7 agar) should also be inoculated.

4. Process additional specimen for additional studies as ordered (*Chlamydia,* mycobacteria, etc.)
5. Spread a drop onto a slide as for a peripheral blood differential, with some very thin areas. Allow to air dry, fix with methanol for 1 minute, and Gram stain.
6. Incubate the anaerobic plates for 7 days before discarding as negative. Hold other plates and process as for routine cultures from the genital tract (look for *Neisseria gonorrhoeae*). Examine thioglycolate broth and subculture only if there is no growth on primary plates.
7. Identify all colony morphotypes and perform susceptibility studies as appropriate. (See "Vaginal Flora Procedure," below.)

VAGINAL FLORA

Routine cultures of vaginal secretions for identification of the predominant microflora are not useful clinically. For diagnosis of bacterial vaginosis, a Gram stain is more reliable than culture,[141,166] even though anaerobic bacteria are involved in this syndrome. Quantitative bacteriological studies of vaginal secretion are important, however, for certain investigational studies.[71,165,260] We recommend two methods, described below.

Vaginal wash [220]

1. In an anaerobic chamber, add a suspension of reduced saline containing 0.02% dithiothreitol (Sigma) and 0.0001% resazurin to sterile 7 ml red top Vacutainer tubes (Becton-Dickinson).

2. Wipe the Vacutainer tube stopper with 70% alcohol and withdraw 3.5 ml of saline into a 5 ml syringe through a large bore needle.
3. Obtain the specimen from the patient during speculum-assisted examination. Use only warm water to lubricate the speculum.
4. Inject the saline into the vagina against the walls (needle removed from syringe). Use a cotton swab to sweep vaginal secretions into the saline pool in the posterior fornix, avoiding the cervix.
5. Use the same syringe attached to a sterile pipette to re-aspirate the washings.
6. Expel all air from the syringe, attach a 20 gauge needle, and inject entire contents into an empty red top tube or anaerobic transport vial.
7. In the anaerobic chamber, prepare serial dilutions of the saline suspension for plating onto appropriate media.
8. Remaining saline suspension may be used for short chain fatty acid analysis.[220]

Calibrated loop[148]

1. In an anaerobic chamber, prepare screwcap tubes containing 9.9 ml of sterile, pre-reduced brain heart infusion broth.
2. Obtain the specimen from the patient during speculum-assisted examination. Use only warm water to lubricate the speculum.
3. Use a 0.01 ml calibrated loop (platinum or plastic disposable) to collect a sample from the pooled secretions in the posterior fornix of the vagina. Insert the loop as vertically as possible.
4. Maintain the brain heart infusion tube under a constant flow of oxygen-free gas (nitrogen is best), open the cap, and quickly inoculate the broth with the loopful of organisms. Immediately retighten the cap.
5. In the anaerobic chamber, vortex, and then prepare serial dilutions of the broth suspension for plating onto appropriate media.

Processing vaginal specimens. However the vaginal secretions are obtained, they should be diluted serially and plated onto a variety of media. Table A-2 suggests media for culturing vaginal wash. Colony forming units usually range up to 10^9/ml when the vaginal wash technique is used. Other methods may yield different quantities and dilutions will need to be adjusted. Rota-Plate (Fisher) is a handy tool for evenly spreading a 0.1 ml inoculum. The Spiral Gradient instrument (Spiral Systems) is also very satisfactory. The Spiral Gradient system reduces the number of dilutions necessary to inoculate.

For detection of clostridia, 1.0 ml of the suspension may be added to 1.0 ml of ethanol, allowed to stand for 45 minutes, and then 0.1 ml can be inoculated to egg yolk agar as described for other plates. Note that a 1:2 dilution of the suspension requires that the final CFU count be doubled. *Gardnerella vaginalis* is most easily detected on human blood bilayer Tween agar plates, on which it appears as small, translucent, beta-hemolytic colonies.[241] Since this organism may be picked from anaerobic plates, all aerotolerance tests on organisms resembling *Gardnerella* (catalase-negative, oxidase-negative, small gram-variable rods) should be carried out on human blood as well as chocolate agar.

All plates should be incubated anaerobically for at least 7 days before final examination to allow slow-growing bacteria (*Actinomyces, Propionibacterium propionicum*) to develop and to allow certain *Porphyromonas asaccharolytica* strains (which may be inhibited on KVLB) to show pigment. *Bacteroides ureolyticus* also may require prolonged incu-

Table A-2. Suggested media for vaginal flora studies[1]

Medium	Incubation atmosphere	Dilutions plated[2]	Days of incubation	Purpose
Brucella PRAS blood agar	Anaerobic	10^{-2}, 10^{-4}, 10^{-6}, 10^{-7}	4-7	Total counts and predominant flora
BBE agar	Anaerobic	10^{-2}, 10^{-4}	2-7	*B. fragilis* group, *Bilophila*
KVLB agar	Anaerobic	10^{-2}, 10^{-4}, 10^{-6}, 10^{-7}	2-7	Bacteroides, pigmented *Prevotella*
BL and BS agar	Anaerobic	10^{-2}, 10^{-4}, 10^{-6}	2-7	*Bifidobacterium*
Lactobacillus selective medium	Anaerobic	10^{-2}, 10^{-4}, 10^{-6}	2-3	*Lactobacillus*
Egg yolk agar (heated or ethanol treated dilutions)	Anaerobic	10^{-2}, 10^{-4}	2-3	*Clostridium*
Rlk or SA agars[3]	Anaerobic	10^{-2}, 10^{-4}, 10^{-6}	4-7	*Mobiluncus*
HV agar[4]	Air+10% CO_2	10^{-2}, 10^{-4}, 10^{-6}, 10^{-7}	2-3	*G. vaginalis*
Chocolate agar	10% CO_2	10^{-2}, 10^{-4}, 10^{-6}, 10^{-7}	2-3	*Haemophilus, Neisseria,* and other capnophiles
Thayer-Martin agar	10% CO_2	10^{-2}	1-2	*N. gonorrhoeae*
Brucella or trypticase soy blood agar	10% CO_2	10^{-2}, 10^{-4}, 10^{-6}	2-3	Total counts and predominant capnophilic flora
MacConkey agar	Air	10^{-2}, 10^{-4}	1-2	Enterics, *Pseudomonas*
BEA[5] agar	Air	10^{-2}, 10^{-4}	1-2	Group D *Streptococcus*
Mannitol salt agar	Air	10^{-2}, 10^{-4}	1-2	*Staphylococcus*
Sabouraud agar	Air	10^{-2}, 10^{-4}	2-7	Yeasts and fungi

[1] See Appendix A for interpretive information on media.

[2] 0.1 ml of each dilution is placed on each medium used.

[3] See reference 216.

[4] HBT agar has also been shown to be a satisfactory medium for the recovery of *G. vaginalis*. See reference 241. (HV and HTB are available from Remel and other sources.)

[5] Bacto Bile Esculin Azide

bation. Its colonies are small, translucent or transparent, and may pit the agar. BBE should be held for 7 days; after the first 48 hours, however, only *Bilophila* species should be sought. These colonies appear small (around 1.0 mm diameter), translucent, usually with black centers, and catalase reaction is strongly positive. "Mobi" agar [221] and Rlk or SA agars [216] have been used for isolation of *Mobiluncus* sp. Colonies may take up to 5 days to appear; they are small and translucent. Descriptions of typical colonies on selective agars are found at the end of this Appendix ("Descriptions of Typical Colonies on Selective Media for Anaerobes").

INTESTINAL FLORA - QUANTITATIVE STUDIES

Collection and Transport

Fecal specimens.   The specimen should be passed naturally and the **entire** specimen collected into a plastic container. A small amount of blood on the outside of the specimen (from rectal bleeding) is allowable. If large amounts are present, the specimen cannot be considered for indigenous flora studies. Specimens contaminated with urine or menstrual flow are to be rejected.

The specimen container should be placed immediately with top loosened into an anaerobic chamber or an anaerobic jar if the specimen is to be transported. Total flora specimens should not be frozen as freezing kills many vegetative cells. Specimens should be maintained at room temperature no longer than 5 hours if they cannot be processed immediately.

Other specimens of intestinal contents.   Specimens from the small bowel or other areas of the bowel are usually collected by syringe through a tube passed into the area to be sampled. The tube should be passed through the nose rather than the mouth. The specimen should be placed into a gassed-out tube or anaerobic transport tube.

Processing of specimens

The directions that follow are used for fecal specimens. These may be modified as necessary for other specimens of intestinal contents.

1. Weigh the specimen, container, and transporter and subtract the weight of the empty transporter and container from the total weight. Place the transporter (with specimen inside) into the anaerobic chamber.
2. After the specimen is in the chamber, homogenize it as follows:
   a. Place the specimen into a special plastic bag and homogenize in the Stomacher (Tekmar Co.; Model # 3500 accommodates a 3.5 liter sample). This method reduces potential aerosols and makes disposal of the unused specimen easier.
   b. Alternatively, especially for larger specimens, a Waring stainless steel container with screw cover (Eberbach #8520) may be used for specimens ≥60 g. The Eberbach glass container (#8470) is used for stools between 30 and 60 g, and the semimicro jar (Eberbach #8580) is used for solid specimens under 30 g and for liquid specimens under 40 g.
   c. Blend non-liquid specimens six times for 30 seconds each time. Do not add diluent. Intermittent blending is necessary so that a sterile glass rod can be introduced into

the blender between each blending operation to scrape the material from the sides of the container and to place it at the bottom of the vessel. This procedure is not necessary with liquid specimens, for which a continuous 3 minute blending is sufficient.

3. Remove an aliquot of approximately 1 g of the homogenized specimen from the chamber and place it on a planchet (pre-weighed square of aluminum foil or plastic weigh-boat) for weighing and drying. Drying is carried out in a vacuum drying oven (15 inch Hg vacuum, 170-180°F) with calcium chloride over 48 hours. Weigh the specimen again following drying. Determine the wet/dry weight ratio so that counts obtained may be corrected and expressed as number of organisms per gram of dry stool.

4. Place another aliquot of approximately 1 g of the homogenized specimen in a sterile tube and weigh it. Add enough reduced 0.05% yeast extract solution to obtain a 10^{-1} dilution. From this dilution, 10^{-2} to 10^{-9} dilutions are made in reduced 0.05% yeast extract solution. See Table A-3 for appropriate dilutions.

   a. Inoculate the first series onto appropriate anaerobic media in the chamber; then remove the dilution series from the chamber for inoculation of media for facultative and aerobic organisms. See Table A-3 and step 9, below, for suggested media.

   b. Heat the second series at 80°C for 10 minutes and then inoculate onto egg yolk agar for isolation of *Clostridium* (spore-forming) species.

   c. Alternatively, prepare only one series of dilutions. Remove 1.0 ml quantities from each tube and place them into empty tubes which may be heated in a heating block within the chamber. If the ethanol method of spore testing is preferred, place a set of small tubes containing 1.0 ml ethanol into the chamber and add 1.0 ml from each of the primary dilution tubes. Mix the suspension with the ethanol and allow the mixture to stand for 30 to 45 minutes.

   d. Finally, plate the heated or ethanol-treated dilutions on egg yolk agar for isolation of spore-forming organisms. Samples from the untreated dilutions are plated onto anaerobic and aerobic media.

5. Dilutions to be plated on the various media are indicated in Table A-3. Aside from the PRAS blood agar plates, which are prepared in the chamber, all other media are prepared on the bench (aerobic media can be purchased commercially) and brought into the chamber 24 hours prior to use. PRAS blood agar and other PRAS plate media are available commercially (Anaerobe Systems). With a pipet, spread 0.1 ml of each dilution to be plated over the medium indicated. This is done by the rotator-pipet method.[6]

6. Place the plates in jars, seal the jars, remove them from the chamber, and incubate for periods as indicated in Table A-3. If equipment permits, the plates may be incubated in the chamber.

7. If one wishes to culture for *Methanobacterium,* use the following procedure:

   a. Bring an unheated set of dilutions in Hungate-type tubes out of the chamber.

   b. Steam PRAS tubes containing 4.9 ml of *Methanobacterium* medium (Appendix C) to melt the agar.

   c. Place the tubes in a 50°C water bath and add 0.02 ml of 5% PRAS sodium sulfide (this may be stored for extended periods) to each tube just before inoculation.

   d. Use a syringe that is first gassed-out with oxygen-free CO_2 for this purpose and for subsequent inoculation of tubes with 0.1 ml of each of the dilutions to be cultured.

   e. After incubation for 1 to 2 weeks, measure methane gas production by analyzing the head-space gas.[161]

Table A-3. Suggested media for anaerobic bowel flora studies[1]

Medium	Dilutions to be plated[2]		Days of incubation	Purpose
	Feces	Small bowel[3]		
PRAS Brucella blood agar	10^{-7}, 10^{-8}, 10^{-9}	10^{-2}, 10^{-4}, 10^{-6}, 10^{-8}	5-7	Total counts and predominant flora
BBE agar	10^{-6}, 10^{-7}, 10^{-8}		2-7	*B. fragilis* group and *Bilophila*
KVLB agar	10^{-2}, 10^{-4}, 10^{-6}, 10^{-8}	10^{-2}, 10^{-4}, 10^{-6}, 10^{-8}	3-5	*Bacteroides,* pigmented *Prevotella*
BL agar (non-selective)/BS agar	10^{-4}, 10^{-6}, 10^{-7}, 10^{-8},	10^{-2}, 10^{-4}, 10^{-6}, 10^{-8}	2-3	*Bifidobacterium*
Lactobacillus selective medium	10^{-2}, 10^{-4}, 10^{-5}, 10^{-6}	10^{-2}, 10^{-4}, 10^{-6}	2-3	*Lactobacillus*
Rifampin blood agar	10^{-4}, 10^{-6}, 10^{-8}		2-3	*F. mortiferum/varium* and certain *Eubacterium* and *Clostridium* species
CCFA[4]	10^{-2}, 10^{-4}, 10^{-6}, 10^{-8}		2-3	*C. difficile*
Veillonella neomycin agar	10^{-2}, 10^{-4}, 10^{-6}	10^{-2}, 10^{-4}, 10^{-6}	2-3	Veillonella and other gram-negative cocci
JVN agar	10^{-2}, 10^{-4}, 10^{-6}	10^{-2}, 10^{-4}, 10^{-6}	2-3	*Fusobacterium* and *Leptotrichia*
Egg yolk agar (heated or ethanol-treated dilutions)	10^{-2}, 10^{-4}, 10^{-6}		2-3	*Clostridium* sp.
PMS agar	10^{-4}, 10^{-6}, 10^{-8}	10^{-2}, 10^{-4}, 10^{-6}	2-3	*Peptococcus* and *Megasphaera*
PS agar	10^{-4}, 10^{-6}, 10^{-8}	10^{-2}, 10^{-4}, 10^{-6}	2-3	*Peptostreptococcus*

[1] See Appendix A for interpretive information on media.

[2] 0.1 ml of each dilution is placed on each medium used.

[3] Selective media optional. Direct examination of specimen indicates whether selective media may or may not be helpful. If specimen is clear fluid, selective media are probably not necessary.

[4] If very low counts of *C. difficile* are to be detected (for epidemiologic studies) use pre-reduced peptone broth containing $39\mu g$/ml cefoxitin and 0.1% sodium taurocholate (262).

8. Microscopic examinations
    a. Prepare a Gram stain from the 10^{-2} dilution. Count bacterial morphotypes seen in 25 fields.
    b. Make Petroff-Hausser chamber counts from the 10^{-3} dilution. Count the numbers of bacteria seen in 25 fields.
    c. Prepare a darkfield examination as follows: Place a small drop from the 10^{-2} and 10^{-3} dilutions on separate cover slips and place the cover slips on microscope slides. Survey each wet preparation at 400 x magnification and a make a detailed examination at 1,000 x magnification. Note motility and describe each morphology. A minimum of 20 minutes should be spent in the examination.
9. Use the dilutions for setting up aerobic cultures in a manner similar to that described for anaerobes. As a minimum, we recommend the following media: tryptic soy or Brucella blood agar (as a nonselective medium), deoxycholate or MacConkey agar (for gram-negative bacilli), PEA or colistin nalidixic acid agar for gram-positive bacteria, cetrimide agar (for *Pseudomonas*), mannitol salt agar (for *Staphylococcus*), Pfizer Selective *Enterococcus* agar (for group D streptococci), potato flake agar medium with chloramphenicol, 50 μg/ml, (for yeasts and fungi; incubate at room temperature), and Mitis-Salivarius agar (for streptococci other than enterococci).

Alternative processing technique—spiral system. The Spiral Plater Model D (Spiral Systems) may be used for plating fecal samples; it eliminates the need for extensive dilutions. This system dispenses 50 μl of sample onto a 100 mm diameter Petri dish in the pattern of a tight spiral; this results in a concentration gradient from the center to the periphery (high counts to low counts). We recommend an initial suspension of 10^5 CFU/ml to obtain a countable plate; experience with particular samples may dictate a need for more concentrated or more dilute suspensions. Based on an estimate of 10^{12} CFU/g of stool, the following dilution scheme may be used:

1. Make a 1:10 dilution (giving a concentration of 10^{11} CFU/g) using 0.05% reduced yeast extract.
2. Continue with three 100-fold dilutions:

Tube A	1:100	10^9 CFU/g
Tube B	1:100	10^7 CFU/g
Tube C	1:100	10^5 CFU/g

3. Plate the dilution from Tube C (10^5 CFU/g). After incubation, count (with the aid of a colony viewer) the number of colonies of each morphotype.
    Note: The colony viewer is equipped with a fixed, plastic counting grid which divides the 100 mm plate into precisely measured quadrants. The total count of each colony type is easily made as the volume of sample dispensed is known for the given areas counted.

Our experience demonstrates that the Spiral Plater is less time-consuming and requires less media than the serial dilution technique. This system can be modified to accommodate 150 mm diameter plates as well. Certain companies (e.g., Anaerobe Systems) offer a variety of media in plates with a level surface (required for this procedure) and will manufacture special plates upon request.

Identification of isolates

After appropriate incubation periods, count the colonies and pick each morphotype to a blood agar plate or other appropriate medium to obtain a pure culture. Descriptions of typical colonies on selective agars used for fecal flora studies are found at the end of this

Appendix (see below). Perform aerotolerance testing and Gram stain from the same colony. PRAS biochemicals and GLC are usually necessary to identify indigenous flora isolates. The total count is obtained by counting all colonies on the non-selective blood agar plate. The total count is multiplied by the wet/dry weight ratio and expressed as CFU/gram of stool, dry weight.

Alternatively, if species level of identification is not desired, characteristics such as Gram stain, disk identification reactions, growth on a specific selective medium, spore test, bile test, and end-product analysis from a PYG or chopped meat broth may give sufficient information to place an organism in a particular group (e.g., *Clostridium* sp., *B. fragilis* group, bile-sensitive gram-negative anaerobic bacilli, anaerobic gram-positive cocci, etc.). We have used this abbreviated identification scheme when looking for gross effects of antimicrobial agents on fecal flora.[178] We also stock all the strains so that further identification may be done later if desired.

DESCRIPTIONS OF TYPICAL COLONIES ON SELECTIVE MEDIA FOR ANAEROBES

As with selective media for aerobes, selective media for anaerobes are not 100% inhibitory toward unwanted organisms; these may grow to a greater or lesser extent. Antibiotics in media, particularly cephalosporins, will deteriorate with time, especially if not stored in the refrigerator, thus allowing growth of unwanted organisms. If selective media are incubated for prolonged periods, unwanted organisms may start to grow. In general, colonies less than or equal to 0.5 mm in diameter at 48 hours should not be picked unless the typical colony morphology of the organism is that size (e.g., *Veillonella* and some *Peptostreptococcus* sp.)

1. Bacteroides bile esculin agar (BBE)
   Colonies of members of the *B. fragilis* group are >1 mm in diameter, circular, entire, raised, and appear gray-brown with brown to black halos. The one exception is *B. vulgatus,* which usually does not hydrolyze esculin and hence appears light gray. Occasional colonies of *Fusobacterium mortiferum* and *F. varium* that may grow on BBE agar are also >1 mm in diameter, but they are flat and irregular. Typical *Bilophila* colonies on BBE are convex with black centers and translucent margins. This medium is highly selective, although occasionally yeast may grow; these appear as dry crinkly colonies and may be dark or gray in color. Also, rare strains of gentamicin-resistant Enterobacteriaceae, enterococci, staphylococci, or *Pseudomonas* may grow; their colony appearance is not predictable, but they are usually less than 1 mm in diameter.
2. *Bacteroides gracilis* selective agar (BGSA)
   Colonies of *B. gracilis* are 1.0 to 2.5 mm in diameter, yellow to brown, translucent, entire, circular and raised. *B. gracilis* does not pit or corrode the BGSA agar.
3. Blood and Liver (BL) and *Bifidobacterium* Selective (BS) agars
   Members of the genus *Bifidobacterium* appear as circular, tan to brown, medium-sized colonies. Even on the selective BS agar, *B. fragilis* group, *Clostridium* sp., and *Lactobacillus* sp. have been isolated from stool specimens. BL agar is not selective but it is an excellent medium for *Bifidobacterium*. All tan to brown colonies should be worked up as possible *Bifidobacterium* sp. We have noted better recovery of *Bifidobacterium* from BL agar than from BS agar.

4. Cadmium sulfate-fluoride-acridine trypticase agar (CFAT)
   Colonies of *Actinomyces* sp. are 1.0 to 3.0 mm in diameter, cream to slightly greenish in color, entire, convex or raised, and opaque. This medium has been formulated for enhanced isolation of *A. naeslundii* and *A. viscosus* from mixed cultures.

5. Cycloserine cefoxitin fructose agar (CCFA)
   C. difficile appears as large yellow colonies with an irregular edge. A few other species of *Clostridium* grow on this medium but the colonies appear more butyrous. The chartreuse fluorescence (under UV light) typical of *C. difficile* is best demonstrated on subculture to a blood agar plate, since the CCFA medium itself is fluorescent. This medium deteriorates rapidly if not refrigerated [65] and works best when fresh. Yeast, enterococci, and other enteric organisms will grow as small flat colonies. Commercial plates vary in their performance characteristics.

6. Egg yolk agar (EYA)
   The numerous species of clostridia have a variety of colony morphologies. *C. perfringens* is lecithinase-positive (white precipitate around the colony). Other *Clostridium* species produce lipase, which appears as a "mother-of-pearl" iridescence on and around the colony. Others may be both lipase- and lecithinase-negative and may not have distinctive colony morphology. It should be noted that for this medium to be selective for *Clostridium* sp., the dilution must first be ethanol- or heat-treated to kill vegetative cells.

7. *Fusobacterium* selective medium (JVN) and *Fusobacterium* neomycin-vancomycin (NV) agar
   Colonies of all species of *Fusobacterium* and *Leptotrichia buccalis* grow on JVN agar [26] (josamycin, vancomycin, and norfloxacin) and NV agar in Fastidious Anaerobe Agar Base and their appearance is similar to that on non-selective blood agar medium. Most other anaerobes are inhibited completely on JVN agar, or their colony size is greatly reduced; rare strains of members of the *B. fragilis* group are only slightly inhibited. Most facultative organisms are completely inhibited on JVN and NV agars.

8. Kanamycin-vancomycin laked blood agar (KVLB)
   Non-pigmented *Bacteroides* and *Prevotella* grow on this medium and may be low convex or large and mucoid and vary in color from gray to yellow to tan to white. The pigmented *Bacteroides* and *Prevotella* generally also grow on this medium. They may have a light tan color initially and become brown to black with prolonged incubation. *Fusobacterium* may occasionally grow on KVLB, especially in mixed culture. *Porphyromonas* sp. is usually inhibited by the vancomycin in the medium. Occasional strains of enterics and yeast may grow on this medium, but their colony morphology is not predictable. Since prolonged incubation may be desirable for pigment production, other organisms will probably grow.

9. *Lactobacillus* selective medium (LBS)
   This is a highly selective medium for a few species of lactobacilli. They may appear as white to gray to yellow to orange in color, and are usually low convex and less than 2 to 3 mm in diameter. One should not assume that this medium will grow all lactobacilli.

10. *Peptococcaceae* and *Megasphaera* selective (PMS) and *Peptostreptococcaceae* selective (PS) agars
    These media, recommended by Ueno and others in Japan, are selective for *Peptostreptococcus* and, in the case of PMS, *Megasphaera* also. These media are available from Eiken Chemical Co.

11. Rifampin blood agar (RIF)

    C. ramosum typically appears as a flat, grayish green, 2-3 mm diameter colony. Isolates of *Eubacterium* may have a similar appearance. *Fusobacterium varium* has a "fried egg" colony and *F. mortiferum* has low convex, tan, translucent colonies.

12. *Veillonella* neomycin agar

    Veillonella appear as small transparent colonies. *Bacteroides* and *Fusobacterium* also grow on this medium and their appearance is similar to that on non-selective blood agar. Most gram-positive and many facultative gram-negative rods are inhibited.

Biochemical Test Procedures

Alkaline Phosphatase (see Glucosidase test)

Bile

The ability of an organism to grow in the presence of 20% bile (equal to 2% oxgall) can be tested with disks impregnated with 20% bile,[251] in liquid medium containing 20% bile, or on BBE agar.

Performance of test

(1) Disk test: Make a fresh subculture of the organism to be tested onto a supportive agar and place a bile disk in the heavy inoculum area. Incubate anaerobically for 24 to 48 hours. Disks are available from Remel and Oxoid.
(2) Broth test: Inoculate two tubes of liquid medium (PRAS PYG or thioglycolate), one supplemented with 2% oxgall and 0.1% sodium deoxycholate and one unsupplemented, with two drops of actively growing culture of the organism. Incubate both tubes until the control tube (unsupplemented) shows good growth.
(3) BBE test: 24 to 72 hour culture on BBE agar.

Interpretation of results

(1) Sensitive (S):
    (a) Disk test: any zone of inhibition present around the bile disk.
    (b) Broth test: less growth in bile-containing tube than in control (not necessarily total inhibition).
    (c) BBE: no growth on BBE agar.
(2) Resistant (R):
    (a) Disk test: no zone of inhibition present around bile disk.
    (b) Broth test: equal or greater (stimulated) growth in bile-containing tube than in control tube.
    (c) BBE: growth on BBE agar; examine for presence of colonies and not simply the blackening of the agar, which may be due to esculin hydrolysis by preformed enzymes present in the inoculum.

Carbohydrate fermentation

See "PRAS Biochemical Inoculation" for inoculation procedure.

Performance of test

(1) Invert the tubes to resuspend settled organisms and record the turbidity (1+ to 4+). At least 2+ growth should be observed to ensure reliable results.
(2) Measure the pH of each tube using a pH meter.

Interpretation of results

(1) Positive: pH below 5.5 = "acid," pH 5.6 - 5.8 = "weak acid."
Note: The pH of PY-base should be above 6.2.
(2) Negative: pH 5.9 and above.

Catalase test

For catalase testing of anaerobic bacteria, 15% H_2O_2 appears to be more sensitive than 3% H_2O_2.

Performance of test

(1) Use 24-72 hour growth on primary plates or on subculture plates, preferably from a medium that does not contain blood. Red blood cells contain catalase and may produce false positive results. Carbohydrate-containing media may also suppress catalase activity.[259]
(2) Touch center of a pure colony with a loop or sterile wooden stick and transfer onto the surface of a clean, dry glass slide or Petri dish. If you must test growth from blood-containing medium, avoid touching the agar.
(3) Add one drop of 15% hydrogen peroxide onto the smear.
   (a) Do not introduce a metallic loop into the drop, because this often causes a false-positive reaction.
(4) Observe for immediate bubbling.
(5) An alternative is to apply a drop of 15% hydrogen peroxide on growth on a medium that does not contain blood and observe for bubble formation.

Interpretation of results

(1) Positive reaction: Immediate and sustained bubbling of the hydrogen peroxide.
(2) Negative reaction: No bubbling observed. Formation of rare bubbles after 20 to 30 seconds is considered a negative catalase test; some bacteria may possess enzymes other than catalase that can decompose hydrogen peroxide.

Conversion of lactate to propionate

Performance of test

(1) Extract an actively growing PY-lactate or chopped meat medium tube (test medium) and a PY-base tube (control) for volatile fatty acid determination by gas chromatography.

(2) Compare the amounts of propionic acid produced.

Interpretation of results

(1) Positive: more propionic acid detected in the test medium than in the PY-base medium.
(2) Negative: equal or less amount of propionic acid produced in test medium compared to PY.

Conversion of threonine to propionate

Performance of test

(1) Extract an actively growing PY-threonine tube (test medium) and a PY-base tube (control) for volatile fatty acid determination by gas chromatography.
(2) Compare the amounts of propionic acid produced.

Interpretation of results

(1) Positive: more propionic acid detected in the test medium than in the PY-base medium.
(2) Negative: equal or less amount of propionic acid produced in the test medium compared to PY.

Desulfoviridin test

Desulfomonas sp. and most members of *Desulfovibrio* sp. contain desulfoviridin pigment. The sirohydrochlorin chromophore of this pigment gives a characteristic red fluorescence when inspected under UV light immediately after addition of 2 N NaOH to a cell suspension.[111]

Performance of test

(1) Inoculate a tube of liquid medium supplemented with 1% pyruvate and 0.25% magnesium sulfate. Incubate until turbidity indicates good growth.
(2) Centrifuge tube to obtain a pellet.
(3) Pipet a heavy drop of the pellet onto a slide.
(4) Add a drop of 2 N NaOH on the pellet and immediately observe under long-wave UV (366 nm) light.

Interpretation of results

(1) Positive: Red fluorescence.
(2) Negative: No fluorescence.

Esculin hydrolysis

The esculin molecule is hydrolyzed to glucose and esculetin by a bacterial enzyme. Released esculetin then reacts with iron salt (ferric ammonium citrate) to form a dark brown or black complex, indicating a positive result. However, H_2S, which is produced by several organisms during metabolism, also reacts with iron to produce a black complex, which interferes with the interpretation of the esculin hydrolysis test. Therefore, all tubes showing darkening after the addition of the reagent need to be checked under UV light; intact esculin fluoresces white-blue under 366 nm light, whereas hydrolyzed esculin has lost its fluorescence.

Performance of test

(1) Add five drops of 1% ferric ammonium citrate solution to a tube of actively growing organism in peptone yeast (PY) esculin broth.
(2) Observe for a color change and fluorescence under UV light.

Interpretation of results

(1) Positive: Black or dark brown color development and no fluorescence under UV light (366 nm).
(2) Negative: No color development or positive fluorescence under UV light.

Flagella stain

The flagella stain described here is that of Ryu, modified by Kodaka *et al.*[138]

Performance of test

(1) Use 48 to 72 hour culture on anaerobic blood agar; CDC blood agar is recommended.
(2) Filter distilled water through a 0.45 μm pore-size filter (for step 3).
(3) Drop a medium size drop of the water on a slide and spread it to the size of a nickel using the pipette.
     Note: do not use slides cleaned with methanol or ethanol.
(4) Pick a colony with an inoculating needle, taking care not to pick up agar, and lightly touch the needle to the center of the drop.
(5) Let the preparations dry in air at ambient temperature.
(6) Make working stain solution: mix 10 parts of solution 1 with 1 part of solution 2 (see Appendix C, Reagents) and filter through glass wool twice.
(7) Flood slides with stain solution and stain for 4-5 minutes.
(8) Wash off majority of stain, then, with strong stream of tapwater, wash the front and the back of the slide (10-20 seconds).
(9) Air-dry the slide and examine microscopically under oil immersion lens beginning at the periphery. You may need to search quite a bit to find cells with intact flagella.

Interpretation of results

(1) Positive: Flagella seen attached to the bacterial cells.
(2) Negative: No flagella seen.

Fluorescence

The use of laked blood or rabbit blood in the medium has been found to enhance the detection of fluorescence of the black/brown pigmented *Porphyromonas* sp. and *Prevotella* sp.

(1) Use a Woods lamp to expose culture plate, patient sample or
infected site to UV light.
   (a) Especially when examining colonies, keep moving the plate till you find the
       optimal angle. You may need to hold the colonies very close to the UV light
       source (within an inch or so), especially if the bulb is weak.
   (b) Fluorescence may take a few seconds to develop.
(2) Note the presence and color of fluorescence.
   (a) It often is necessary to reincubate plates for several days before the fluorescence
       is visible.

Interpretation of results (Table B-1)

(1) Positive: Distinct brick red color detected with Woods lamp
    **Note: Brick red fluorescence is the only reliable color for presumptive identifi-
    cation of *Porphyromonas* sp. and pigmented *Prevotella* sp. (See Figure 4-7).
    Other anaerobes may also have characteristic fluorescence (see Table B-1).**
(2) Negative: No color change detected.

Table B-1. Fluorescence of selected anaerobes

Organism	Color
Porphyromonas asaccharolytica-endodontalis	Red, orange
P. gingivalis	No fluorescence
Pigmented *Prevotella* sp.	Red
Non-pigmented gram-negative bacilli	No fluorescence or pink orange, yellow
Fusobacterium sp.	Chartreuse
Veillonella sp.	Red
Eubacterium lentum	Red or no fluorescence
Clostridium difficile	Chartreuse
C. innocuum	Chartreuse
C. ramosum	Red

Gelatin liquefaction

Performance of test

(1) Refrigerate an actively growing ($>2+$ turbidity) PRAS gelatin tube along with an
    uninoculated tube (negative control) for at least one hour.
(2) Remove tubes to room temperature and invert immediately. Observe the test and
    control tubes for liquefaction every 5 minutes (the solidified agar begins to melt and
    falls toward the inverted top of the tube). Record results.

(1) Positive: Gelatin medium in the inoculated tube fails to solidify (drops to the top of the inverted tube immediately).
(2) Weak positive: The inoculated tube becomes liquid when it reaches room temperature ($<$30 min).
(3) Negative: tube fails to liquefy when it reaches room temperature ($>$30 min).

Glucosidase test (α-fucosidase, α-glucosidase, β-N-acetylglucosaminidase, ONPG, β-glucuronidase, β-xylosidase) and alkaline phosphatase test

The presence of chromogenic glycosidases or alkaline phosphatase enzymes can be detected in commercial identification kits, such as RapID ANA, Rapid ID 32A or API ZYM, or separately using Rosco diagnostic tablets (described below).

Performance of test

(1) Use actively growing (48 to 72 hour) culture on an agar plate medium that does not contain carbohydrates.
(2) Make a heavy suspension of organism ($>$2.0 McFarland turbidity) in 0.25 ml of saline.
(3) Add the appropriate Rosco tablet, vortex the suspension, and incubate 37°C for 4 hours.

Interpretation of results

(1) Positive: Yellow color development. When testing β-N-acetylglucosaminidase, only a strong yellow color should be recorded as positive.
(2) Negative: No color or very pale yellow color.

Growth stimulation test

Growth supplements are required for growth or for stimulation of growth by certain anaerobes. If the organism fails to grow in a medium such as PYG or thioglycolate, supplements should be added.

(1) 0.5% arginine is required by *E. lentum*.
(2) Formate-fumarate (0.3% and 0.3%) is required by *B. ureolyticus* group.
(3) 1% pyruvate stimulates the growth of *Bilophila* and *Veillonella* sp.
(4) 1% pyruvate with 0.25% magnesium sulfate stimulates sulfate-reducing bacteria.
(5) N-acetylmuramic acid (0.001%) stimulates *B. forsythus*.
(6) Hemin stimulates pigmented gram-negative rods and the *B. fragilis* group.
(7) Sodium bicarbonate (1%) stimulates many gram-negative rods.
(8) 0.5% Tween-80 enhances the growth of gram-positive organisms.
(9) Serum (usually add 1%) can be used to stimulate the growth of gram-positive and gram-negative anaerobes.

Performance of test. Add supplement(s) to appropriate concentration and incubate for 48 to 72 hours along with a tube without the supplement(s) as a control.

Interpretation of results. Compare the growth of supplemented tubes and the control tube; if growth is substantially better (more turbid) in the presence of supplement, that is interpreted as "positive." Add the supplement providing the best enhancement to all the tubes inoculated for biochemical tests (i.e., PRAS biochemicals, motility).

Indole production (tryptophanase)

To perform the indole test, growth medium that contains tryptophan is mandatory [229]. Also, since the enzyme that degrades tryptophan is diffusible in agar, it is important to have only one culture of an organism per plate. Do not use a plate that has a nitrate disk on it; nitrate can interfere with the spot indole test by inducing false-negative results. It also has been reported that false-negative results may be obtained in indole-nitrate broth medium when nitrite concentrations reach 0.25 to 0.75 mg/ml (nitrate test positive) [218].

Performance of test

(1) Spot test:
    (a) Inoculate an agar medium that contains sufficient tryptophan, such as Brucella blood agar or egg yolk agar.
    (b) Place a piece of filter paper (No. 1 Whatman) in a clean petri dish cover.
    (c) Moisten the filter paper with paradimethylaminocinnamaldehyde reagent. The paper should be saturated, but not "dripping" wet.
    (d) Remove several colonies from the agar with a wooden stick or loop and rub on the filter paper. When testing anaerobic bacteria, it is good to use a heavy inoculum, since sometimes the reaction can be fairly weak. Several tests may be performed on a single filter paper, but once it dries, it must be replaced.
    (e) Alternatively, a sterile paper disk can be placed on a heavy growth area for 5 minutes, then removed to an empty petri dish and a drop of reagent added to the disk.
(2) Broth test:
    (a) Inoculate organism into indole-nitrate medium or chopped meat broth. Incubate until good growth occurs.
    (b) Remove 2 ml of culture to a small tube for testing.
    (c) Add 1 ml of xylene, stopper with a rubber stopper, shake vigorously, and let stand for 15 minutes with the cap on. You may wish to crack the cap to release gas pressure.
    (d) Add 0.5 ml of Ehrlich reagent slowly down the side of the tube.

Interpretation of results

(1) Positive:
    (a) Spot test: development of a blue or green color around the inoculum within 30 seconds. The dark-pigmented organisms can produce a greenish color, or sometimes the color development may be "masked" by the pigment and therefore the inoculum must be examined with care (see Figure 4-8).
    (b) Broth test: development of pink or fuchsia ring in a layer directly under the Ehrlich's reagent within 15 min.

(2) Negative:
   (a) Spot test: no color change or pinkish color. Late color development should be disregarded.
   (b) Broth test: development of a yellow ring. Late weak color development should be disregarded.

Lecithinase

Lecithinase mediates the breakdown of lecithin to diglyceride and phosphorylcholine. On EYA, the formation of insoluble diglycerides causes a visible opacity around the colony.

Performance of test.   Subculture the organism onto EYA and incubate for 24 to 72 hours.

Interpretation of results

(1) Positive: white, opaque, diffuse zone and extending into the medium surrounding the colonies (see Figure 4-10).
(2) Negative: no reaction on the agar. Compare negative plate with uninoculated plate, since lecithinase can diffuse throughout the whole agar and make interpretation difficult.

Lipase

Bacterial lipases hydrolyze the breakdown of triglycerides into glycerol and free fatty acids. Fatty acids are mostly insoluble and cause opacity on EYA, producing an iridescent sheen on the colonies and the surface of EYA. Unlike lecithinase, lipase is not diffusible, and the reaction occurs only on the surface of the agar in the immediate vicinity of the colony.

Performance of test.   Subculture the organism to EYA and incubate for 24h to 1 week.

Interpretation of results

(1) Positive: an iridescent sheen on the surface of bacterial growth and on the agar surface around the colonies (observe under oblique light) (see Figure 4-11). The lipase reaction can take up to one week to occur for some isolates.
(2) Negative: no reaction on the agar.

Milk

Anaerobic bacteria may produce four different reactions in milk medium: acid, clotting, gas, and digestion.

Performance of test

(1) Follow same procedure to inoculate PRAS milk tube as is described for inoculation of PRAS biochemical tests.

(2) Incubate tube at 35°C for 4 days to 3 weeks.
(3) Observe for presence of a) acid, b) gas, c) clot, d) digestion, and record results.

Interpretation of results

(1) Acid pH: lactose fermentation; detected with a pH electrode and performed at the end of the incubation period.
(2) Gas: presence of bubbles, release of pressure upon opening cap, or stormy fermentation (gas disrupting a clot).
(3) Clot: either lactose fermentation (coagulation of casein in low pH milieu) or hydrolysis of casein by a rennin-like enzyme.
(4) Digestion: clearing of the milk in the tube; can take 4d to 3 weeks, may occur with curd formation or the curd may be digested. Be sure to differentiate digestion from the whey that is also produced. If the pH is strongly acid after incubation, then digestion has probably not taken place.

Motility

Prepare a wet mount or use the "hanging drop" technique to examine an actively growing culture in a liquid medium. Look for directed movement that can be differentiated from both Brownian movement and the flow of organisms in streams through the medium.

Nagler test

Clostridium perfringens type A antitoxin inhibits the production of lecithinase of *C. perfringens* on EYA plate. Other species that will give positive Nagler reaction are *C. bifermentans, C. sordellii,* and *C. baratii.*

Performance of test

(1) Swab one half of an EYA plate with *C. perfringens* type A antitoxin and allow it to dry.
(2) Make a single streak of the organism starting from the side of the EYA plate that does not contain antitoxin.
(3) Incubate for 24 to 48 hours.

Interpretation of results

(1) Positive: Positive lecithinase reaction on the half of the EYA plate without antitoxin and inhibition of lecithinase reaction on the half containing the antitoxin (see Figure 4-13).
(2) Negative: Lecithinase reaction not inhibited on either side of the plate *or* no reaction on the agar.

Nitrate reduction test

Performance of test

(1) Disk test:
    (a) Make a fresh subculture of the organism to be tested onto a supportive agar and place a nitrate disk (available commercially) in the heavy inoculum area. Incubate anaerobically for 24 to 48 hours.

(b) Remove the disk from the surface of the plate and place it in a clean petri dish or
    on a slide.
(c) Add 1 drop each of reagents A and B (see Appendix C). If no color develops in a
    few minutes, drop a small amount of zinc dust or zinc granules onto the surface of
    the disk and observe up to five minutes.

(2) Broth test:
    (a) Inoculate organism into indole-nitrate medium. Incubate until good growth
        occurs.
    (b) Remove 1 ml of culture to a small tube for testing.
    (c) Add 0.2 ml each of reagents A and B. If no color develops in a few minutes, add a
        small amount of zinc dust or zinc granules to the tube and observe for five
        minutes.

Interpretation of results

(1) Positive: red or pink color development after adding the reagents *or* no color develop-
    ment after adding zinc (see Figure 4-9).
(2) Negative: no color development after adding the reagents **and** red color development
    after adding zinc.

PRAS biochemical inoculation

Performance of test

(1) Use an actively growing broth culture (without carbohydrate) of at least 2+ turbidity.
    Invert the culture to provide an even suspension, and inoculate two to ten drops into
    the appropriate PRAS biochemicals including a PY-base. Inoculation should be
    done under anaerobic atmosphere (in a chamber or under continuous gas flow into
    the tube during manipulation), or through a rubber stopper using a needle and
    syringe.
(2) If the organism grows poorly, be certain to add the necessary and appropriate
    supplements to each tube (see "Growth stimulation test," above). Tween and hemin
    can be added directly to the inoculum broth.
(3) After the last tube has been inoculated, plate one drop of the inoculum onto a BA
    plate to check for purity and viability (incubate anaerobically), and one drop onto
    either BA or chocolate agar for CO_2 incubation.
(4) Incubate at 37°C until good growth is seen (>2+). If 2+ growth is not reached after
    one week of incubation, a better supplementation should be sought.
(5) Interpret the carbohydrate fermentation results by measuring acid production using a
    pH meter (see "Carbohydrate fermentation," above), read other biochemical tests,
    and process appropriate tubes for GLC.

Reverse-CAMP test [102]

α-toxin-producing *Clostridium perfringens* and β-hemolytic *Streptococcus agalactiae*
form a characteristic arrowhead-shaped, synergistic hemolytic pattern on blood agar.

Performance of test

(1) Inoculate a blood agar plate with a single streak of suspected *C. perfringens*.
(2) Make a single streak of group β-hemolytic *Streptococcus* at a 90 degree angle to
    within a few mm of the suspected *C. perfringens*.

(3) Incubate for 24 to 48 hours.

Interpretation of results

(1) Positive: Synergistic arrowhead-shaped hemolytic pattern present (see Figure 4-12).
(2) Negative: No synergistic action observed.

Rosco disk identification

See "glycosidase test," "trypsin-like activity," and "urease test."

Special potency disk identification

Performance of test

(1) Subculture the isolate onto a blood agar plate. To ensure an even, heavy lawn of growth, streak the first quadrant back and forth several times. Streak the other quadrants to yield isolated colonies.
(2) Place the antibiotic disks on the first and second quadrants well apart (~20 mm) from each other.
(3) If you have several isolates to test, first streak all the plates and then add the disks to them at the same time.
(4) Incubate the plate(s) anaerobically for 48 to 72 hours at 35-37°C.
(5) Examine for a zone of inhibition of growth surrounding the disk.

Interpretation of results (Table B-2)

(1) Vancomycin, kanamycin, colistin disks:
    (a) Sensitive (S): Zone of inhibition is ≥10 mm.
    (b) Resistant (R): Zone of inhibition is <10 mm.
(2) SPS disk
    (a) Sensitive (S): Zone of inhibition is ≥12 mm. The value of 12 mm is the guideline according to the literature. In practice, *P. anaerobius* usually gives a very large zone (≥16 mm), whereas the other anaerobic cocci that appear to be sensitive to SPS produce a smaller zone. To presumptively identify *P. anaerobius*, examine the isolate for the presence of characteristic Gram and colonial morphology and caramel-like odor.
    (b) Resistant (R): Zone of inhibition is <12 mm.

Spore test (ethanol) [139]

Spores can be observed in Gram stained preparations, under darkfield microscopy, or they can be demonstrated with the spore test. Some commonly encountered *Clostridium* sp., such as *C. perfringens* and *C. ramosum,* sporulate poorly and spores are rarely seen.

Performance of test

(1) Use an actively growing broth culture of at least 2+ turbidity.
(2) Let the culture stand at room temperature for 1 week.

Table B-2. Special potency disk patterns

	Kanamycin 1000 µg	Vancomycin 5 µg	Colistin 10 µg	SPS 1000 µg
Gram-negative	V	R[1]	V	
Gram-positive	V	S[2]	R	
Bacteroides fragilis group	R	R	R	
B. ureolyticus group	S	R	S	
Fusobacterium sp.	S	R	S	
Porphyromonas sp.	R	S	R	
Prevotella sp.	R	R	V	
Veillonella sp.	S	R	S	
Peptostreptococcus anaerobius	R[S]	S	R	S
Other gram-positive cocci	S	S	R	R

[1] *Porphyromonas* sp. is vancomycin-sensitive.

[2] Rare stains of *Lactobacillus* sp. and *Clostridium* sp. may be vancomycin-resistant.

(3) Plate one drop of the culture onto a BA plate for anaerobic incubation to check for purity and viability (pre-spore test), and one drop onto either BA or chocolate agar for CO_2 incubation (purity control).
(4) Add equal volumes of the culture to 95% ethanol and mix gently. Incubate at room temperature for 30 minutes.
(5) Subculture mixture on BA plate for anaerobic incubation (post-spore test).
(6) Incubate plates for 2 days.
(7) Examine for growth.

Interpretation of results

(1) Positive: Growth present on both anaerobic plates (pre- and post-).
(2) Negative: Growth on the pre-spore test plate and no growth on the plate inoculated after the spore test.

Trypsin-like activity

Trypsin-like activity can be demonstrated using several methods including Rosco tablets, API ZYM system, BANA reagent (N-benzoyl-DL-arginine-2-naphthylamide),[143] and CAAM reagent (carbobenzoxy-L-arginine-7-amino-4-methylcoumarin amide-HCl).[212] The method utilizing Rosco tablets is described in this manual.

Performance of test

(1) Use an actively growing (48 to 72 hour) culture on an agar medium that does not contain carbohydrates. Make a heavy suspension of the organism (>2.0 McFarland turbidity) in 0.25 ml of saline.

(2) Add a Rosco trypsin tablet, vortex and incubate at 37°C for 4 to 18 hours.
(3) After incubation, add a drop of Fast Blue BB reagent (Rosco) and wait 5 minutes or examine under long-wave UV light.

Interpretation of results

(1) Positive: salmon color after adding Fast Blue BB reagent or blue fluorescence under UV light.
(2) Negative: yellow (green) color after adding Fast Blue BB reagent or no fluorescence under UV light.

Urease

Performance of test

(1) Rapid test: Make a heavy suspension of the actively growing organism in 0.25 ml of saline and add a urea disk (Difco, Rosco). Alternatively, make a heavy suspension of the organism in 0.5 ml of rapid urea broth (Appendix C). Incubate aerobically at 37°C for 4 to 24 hours.
(2) Broth test:
    (a) Inoculate a PY-base medium and a PY medium supplemented with 2% urea. Incubate until good growth occurs.
    (b) Determine the pH of both the PY and the PY with urea.

Interpretation of results

(1) Positive:
    (a) Rapid test: bright red color in urea broth or urea disk tube.
    (b) Broth test: pH of PY-urea tube at least 0.3 units higher than pH of PY-base.
(2) Negative:
    (a) Rapid test: no color change to yellow color in urea broth or urea disk tube.
    (b) Broth test: ≤ 0.3 unit pH change in PY-urea tube.

Appendix C

Preparation of Media and Reagents

QUALITY CONTROL PROCEDURES

All media, reagents, and test systems should be monitored for acceptable performance. The Wadsworth QC procedures have been established based on the recommendations of the NCCLS Quality Assurance Document M22-A, Vol. 10, No. 14. for commercially prepared microbiological culture media.[182] Quality control procedures for gas-liquid chromatography and antimicrobial susceptibility testing have been described in chapters 7 and 8. Where tests such as nitrate reduction are common to aerobic, facultative and anaerobic bacteria, results obtained for the aerobic or facultative bacteria will suffice.

All anaerobic systems should be tested daily for anaerobiosis using an indicator such as resazurin or methylene blue. When using resazurin, it must be protected from light to prevent inactivation. If the methylene blue strip is not white upon opening the foil packet, discard and use a fresh one.

Quality control procedures suggested for commonly used anaerobic media and reagents are given in Table C-1. Known positive and negative organisms are always used when testing commercially prepared or in-house prepared media and reagents. Each batch or lot number of plated or biochemical test medium should be tested. Antibiotic disks, other disks used for identification, additives for growth stimulation, and indole test reagents should be tested at weekly intervals. The following cultures should be maintained for use in quality control procedures:

Anaerobes

B. fragilis, ATCC 25285	
B. thetaiotaomicron, ATCC 29741	*B. ureolyticus*
P. melaninogenica, ATCC 25845	*F. necrophorum*
C. perfringens, ATCC 13124	*E. lentum*
B. ovatus, ATCC 8483	*P. micros*
P. anaerobius	

Table C-1. Quality control procedures for media and reagents

Media or Reagent	Test Organism	Results/Rationale
AGARS		
Blood agar	*B. fragilis*	growth
	B. levii	growth, pigment [1]
	C. perfringens	growth, double zone of β-hemolysis
	F. nucleatum	growth
	P. anaerobius	growth
	P. melaninogenica	growth, pigment [1]
BBE	*B. fragilis*	growth, black colonies
	P. melaninogenica	no growth, bile inhibition
	E. coli	no growth, gentamicin inhibition
KVLB	*P. melaninogenica*	growth, black pigment [1]
	C. perfringens	no growth, vancomycin inhibition
	E. coli	no growth or strong inhibition by kanamycin
PEA	*B. fragilis*	growth
	P. mirabilis	inhibition of swarming
EYA	*C. perfringens*	growth, lecithinase production
	F. necrophorum	growth, lipase production
CCFA	*C. difficile*	growth, chartreuse fluorescence
	B. fragilis	inhibition of growth by cefoxitin
	E. coli	inhibition of growth by cycloserine
DISKS		
Colistin (10 µg)	*F. necrophorum*	susceptible
	B. fragilis	resistant
Kanamycin (1 mg)	*C. perfringens*	susceptible
	B. fragilis	resistant
Vancomycin (5 µg)	*C. perfringens*	susceptible
	B. fragilis	resistant

Table C-1. Quality control procedures for media and reagents (continued)

Medium or Reagent	Test Organism	Results/Rationale
Nitrate	*B. ureolyticus*	nitrate reduced
	B. fragilis	nitrate not reduced
SPS	*P. anaerobius*	susceptible, zone ≥ 12 mm
	P. micros	resistant, zone < 12 mm
GROWTH STIMULATION TESTS		
Base medium + F/F	*B. ureolyticus*	growth enhanced compared to that in base medium
Base medium + pyruvate	*B. wadsworthia*	growth enhanced compared to that in base medium
Base medium + arginine	*E. lentum*	growth enhanced compared to that in base medium
Base medium + Tween 80	*P. micros*	growth enhanced compared to that in base medium
MISCELLANEOUS		
Urea broth	*B. ureolyticus*	urease produced
	B. fragilis	urease not produced
Indole test reagents	*F. necrophorum*	indole produced
	B. fragilis	indole not produced
Thioglycolate	*B. fragilis*	growth in 24 hours
PY base	*B. fragilis*	growth in 24 hours

[1] pigment production may require more than 48 hours incubation.

Aerobes

E. coli, ATCC 25922
E. faecalis, ATCC 29212

P. mirabilis

Some of these strains are available as freeze-dried disks (Anaerobe Systems, BBL, and American Type Culture Collection). Well-documented laboratory strains may be used where ATCC strains are not listed.

MEDIA AND REAGENTS

Most of the media suggested for use are available in dehydrated form and are prepared according to the directions of the manufacturer. Supplements are listed for each medium when enrichment is advisable or required. Most agar bases may be dispensed before autoclaving into screw-capped tubes or 100 ml bottles and stored until needed. They can then be remelted, supplemented and poured into plates. Many prepared agar media are available from local media suppliers; PRAS blood agar, kanamycin-vancomycin laked blood agar, Bacteroides bile-esculin, egg yolk, and phenylethyl alcohol agar plates are available from Anaerobe Systems. Shelf life of media stored under anaerobic conditions is generally longer than when they are stored in air.

PRAS media used for biochemical tests and fermentation reactions are prepared as described in the VPI Anaerobe Laboratory Manual [109] and are supplemented with vitamin K_1 and hemin; they can be purchased from Carr-Scarborough Microbiologicals, Inc. The use of thioglycolate-based medium recommended by CDC [58] is an acceptable alternative to the use of PRAS media; however, all reactions listed in this manual are based upon results obtained in PRAS media.

Antitoxin - *C. perfringens* type A.   Available from Pittman and Moore, Inc. (for use in the Nagler test).

Arginine (10%)

a. Ingredients

Arginine HCl	10 g
Distilled water	100 ml

b. Preparation
1) Dissolve the arginine HCl in water.
2) Sterilize by filtration.
3) Store in the refrigerator.

c. Use

Add 0.5 ml of arginine solution to 10 ml of freshly steamed thioglycolate medium for growth stimulation of *E. lentum*.

Bacteroides Bile Esculin Agar (BBE)[150]

a. Ingredients

Trypticase soy agar	40 g
Oxgall	20 g
Esculin	1 g
Ferric ammonium citrate	0.5 g
Hemin solution (5 mg/ml)	2 ml
Gentamicin solution (40 mg/ml)	2.5 ml
Distilled water	1000 ml

b. Preparation
1) Combine all the ingredients.
2) Adjust the pH to 7.0.
3) Boil or steam to dissolve.
4) Autoclave at 121°C for 15 minutes.
5) Cool to 50°C and pour plates.

c. Shelf life
    Plates - 2 weeks at room temperature
    3 months at refrigerator temperature
    Bottles - 1 year at refrigerator temperature

Bacteroides gracilis Selective Agar (BGSA)[147]

a. Ingredients
Tryptic Soy Agar (Difco)	8 g
Distilled water	200 ml
Nalidixic acid (1.6 mg/ml)	1 ml
Teicoplanin (6.4 mg/ml)	1 ml
Formate-fumarate additive	10 ml
Potassium nitrate	0.4 g

b. Preparation
    1) Combine agar base and water.
    2) Boil to dissolve.
    3) Autoclave at 121°C for 15 minutes.
    4) Cool to 50°C and add nalidixic acid, teicoplanin, and formate-fumarate, then
        pour plates.
    5) Quantity is sufficient for 10 plates, 20 ml/plate.
c. Shelf life
    2 weeks at refrigerator temperature

BL (Blood and Liver) Agar

a. Ingredients
BL agar (Nissui Pharmaceutical Co.)	58 g
Distilled water	1000 ml
Defibrinated horse blood	50 ml

b. Preparation
    1) Combine agar base and water.
    2) Heat to dissolve.
    3) Autoclave at 115°C for 20 minutes.
    4) Cool to 50°C, add defibrinated horse blood, and pour plates.
c. Shelf life
    2 weeks at refrigerator temperature

BS (*Bifidobacterium* Selective) Agar

a. Ingredients
BL agar (Nissui Pharmaceutical Co.)	58 g
BS solution (Nissui Pharmaceutical Co.)	50 ml
Defibrinated horse blood	50 ml
Distilled water	1000 ml

b. Preparation
    1) Combine agar base and water.
    2) Heat to dissolve.
    3) Autoclave at 115°C for 20 minutes.
    4) Cool to 50°C, add BS solution and defibrinated horse blood, and pour plates.
c. Shelf life
    2 weeks at refrigerator temperature

Brucella Blood Agar

a. Ingredients

Brucella agar (BBL or Difco)	43 g
Hemin solution (5 mg/ml)	1 ml
Vitamin K_1 solution (1 mg/ml)	1 ml
Distilled water	1000 ml
Sterile defibrinated sheep blood	50 ml

b. Preparation
1) Combine all ingredients except sheep blood.
2) Boil to dissolve.
3) Autoclave at 121°C for 15 minutes
4) Cool to 50°C, add sheep blood, and pour plates.

c. Shelf life
Plates - 2 weeks in refrigerator
Bottles - 1 year at refrigerator temperature

Brucella Blood Agar for susceptibility testing

a. Ingredients

Brucella agar (BBL or Difco)	43 g
Vitamin K_1 solution (1 mg/ml)	1 ml
Distilled water	1000 ml
Sterile laked sheep blood	50 ml

Hemin ($5\mu g$/ml) may be added to improve growth of some organisms. Similarly, other supplements may be added to enhance growth of fastidious anaerobes, but care should be taken to ensure that quality control organisms are within range.

b. Preparation
1) Combine all ingredients except laked blood.
2) Boil to dissolve.
3) Autoclave at 121°C for 15 minutes.
4) Media may be refrigerated in bottles or flasks, and used as needed for suscepti- bility tests. Blood should be added when the test plates are poured (see Chapter 8).

c. Shelf life
For media without added blood up to one month.

Brucella Broth. Brucella broth is used in broth dilution susceptibility tests.

a. Ingredients

Brucella broth (BBL or Difco)	28 g
Vitamin K_1 solution (1 mg/ml)	1 ml
Hemin solution (5 mg/ml)	1 ml
Sodium bicarbonate (20 mg/ml)	5 ml
Distilled water	1000 ml

b. Preparation
1) Dissolve ingredients.
2) Dispense in convenient volumes.
3) Autoclave at 121°C for 15 minutes.

c. Shelf life
1 year at refrigerator temperature

Cadmium Sulfate-Fluoride-Acridine Trypticase Agar (CFAT)[268]

a. Ingredients

Trypticase soy broth	30 g
Glucose	5 g
Agar	15 g
Cadmium sulfate	13 g
Sodium fluoride	80 g
Neutral acriflavine	1.20 mg
Basic fuchsin	0.25 mg
Distilled water	1000 ml
Sterile potassium tellurite	2.5 mg
Sterile defibrinated sheep blood	50 ml

b. Preparation
1) Combine all ingredients except potassium tellurite and sheep blood.
2) Boil to dissolve.
3) Autoclave at 121°C for 15 minutes.
4) Cool to 50°C, add potassium tellurite and sheep blood, then pour plates.

c. Shelf life - unknown

Chocolate-Bacitracin Agar (Modified from reference 112.)

a. Ingredients

Blood agar base	40 g
Distilled water	1000 ml
Sterile defibrinated sheep blood	50 ml
Bacitracin	300 mg

b. Preparation
1) Combine blood agar base and water.
2) Boil to dissolve.
3) Autoclave at 121°C for 15 minutes.
4) Cool to 50°C. Add blood.
5) Raise temperature to 70 to 80°C, mixing constantly but gently until blood turns a chocolate color.
6) Cool to 50°C, add bacitracin and pour plates.

c. Shelf life
Plates - 1 week at refrigerator temperature
Bottles - 1 year at refrigerator temperature

Cycloserine-Cefoxitin Fructose Agar (CCFA) [85]

a. Ingredients

Proteose peptone No. 2 (Difco)	40 g
Disodium phosphate	5 g
Monopotassium phosphate	1 g
Fructose	6 g
Agar	20 g
Neutral red (1% in ethanol)	3 g
Distilled water	1000 ml
Cycloserine	500 mg
Cefoxitin	16 mg

b. Preparation
  1) Combine all ingredients except cycloserine and cefoxitin.
  2) Adjust pH to 7.6.
  3) Boil to dissolve.
  4) Autoclave at 121°C for 15 minutes.
  5) Cool to 50°, add cycloserine and cefoxitin, then pour plates.
c. Shelf life
  Plates - 8 weeks at refrigerator temperature
  Bottles - 1 year at refrigerator temperature

Egg Yolk Agar (EYA)

a. Ingredients

Proteose peptone No. 2 (Difco)	40 g
Disodium phosphate	5 g
Monopotassium phosphate	1 g
Sodium chloride	2 g
Magnesium sulfate	0.1 g
Glucose	2 g
Vitamin K_1 (1 mg/ml)	1 ml
Hemin solution (5 mg/ml)	1 ml
Agar	20 g
Distilled water	1000 ml
Egg yolk emulsion (Difco or laboratory preparation, see below)	50 ml
or Egg yolk emulsion (Oxoid)	74 ml

b. Preparation
  1) Combine all ingredients except egg yolk, vitamin K_1, and hemin.
  2) Adjust pH to 7.6.
  3) Boil to dissolve.
  4) Autoclave at 121°C for 15 minutes.
  5) Cool to 50°C; add egg yolk emulsion, vitamin K_1, and hemin, and pour plates.
c. Shelf life
  Plates - 1 week at refrigerator temperature
  Bottles - 1 year at refrigerator temperature

Egg Yolk Emulsion

a. Ingredients
  Fresh eggs
  Sterile normal saline
b. Preparation
  1) Cleanse fresh eggs thoroughly with an alcohol pad.
  2) Punch an airhole in one end and a hole approximately 10 mm in diameter in the other end.
  3) Carefully allow the egg white to drip out, assisting as necessary with sterile syringe and needle.
  4) When white has been removed, aspirate yolk into a sterile syringe.
  5) Place into sterile flask with glass beads and an equal volume of normal saline.
  6) Mix thoroughly and test sterility by plating a drop on Brucella blood agar and inoculating 1 ml into supplemented thioglycolate medium. Prepared egg yolk emulsion is available from Difco and Oxoid.

a. Ingredients
    Para-dimethylaminobenzaldehyde   1 g
    Ethyl alcohol (95%)   95 ml
    Hydrochloric acid (concentrated)   20 ml
b. Preparation
    1) Dissolve para-dimethylaminobenzaldehyde in alcohol.
    2) Slowly add hydrochloric acid.
    3) Store in a dark bottle refrigerated for up to one year.

Ferric Ammonium Citrate (1%)

a. Ingredients
    Ferric ammonium citrate   1 g
    Distilled water   100 ml
b. Preparation
    1) Dissolve ferric ammonium citrate in water.
    2) Store in a dark bottle refrigerated for up to one year.

Fildes Enrichment.   Available from BBL and Difco

Flagella stain (Ryu) (Also available from Carr Scarborough.)

Solution 1:
a. Ingredients
    5% Carbolic acid (Mallinckrodt)   100 ml
    Tannic acid (J.T. Baker Chem. Co)   20 g
    Saturated Aluminum potassium sulfate 12-hydrate   100 ml
    ($\sim$13.2g/100 ml) (J.T. Baker Chem. Co)
b. Preparation
    1) Mix 5% carbolic acid (phenol) solution, tannic acid, and saturated potassium
        sulfate 12-hydrate solution. Use a brown glass bottle.
    2) Store at room temperature.
        **Note: when using carbolic acid crystals, first melt crystals in 50 °C water
        bath, then pipette 5 ml of liquefied carbolic acid into 95 ml of distilled water.
        Use a glass bottle.**

Solution 2:
    Saturated solution of crystal violet in ethanol (12 g per 100 ml). Store at room
    temperature.

Final stain preparation:
    1) Mix 10 parts of solution 1 with 1 part of solution 2 in a brown glass bottle.
    2) Filter the final stain twice thorough glass wool before using.

Formate-Fumarate Additive

a. Ingredients
    Sodium formate   3 g
    Fumaric acid   3 g
    Distilled water   50 ml
    Sodium hydroxide   20 pellets

b. Preparation
   1) Combine ingredients, except for 10 ml of water. Stir until pellets are dissolved and fumaric acid is in solution.
   2) Bring pH to 7.0 with 4N sodium hydroxide and bring final volume to 50 ml with distilled water.
   3) Sterilize by filtration.
c. Use
   Add 0.5 ml of the F/F additive to 10 ml of culture medium to stimulate growth of *B. ureolyticus* and other fastidious gram-negatives.

Fusobacterium Selective (JVN) Agar

a. Ingredients
Fastidious Anaerobe Agar (FAA) (Lab M)	46 g
Josamycin (Yamanouchi Pharmaceuticals)	3 mg
Vancomycin	4 mg
Norfloxacin	1 mg
Distilled water	1000 ml
Sterile defibrinated horse blood	50 ml
b. Preparation
   1) Combine agar base and distilled water.
   2) Bring to boil to dissolve.
   3) Autoclave at 121°C for 15 minutes.
   4) Cool to 50°C, add blood, vancomycin, norfloxacin, and josamycin, then pour plates.
c. Shelf life
   Plates - 2 weeks at refrigerator temperature
   Bottles - 1 year at refrigerator temperature

Fusobacterium Neomycin-Vancomycin (NV) Agar

a. Ingredients
Fastidious Anaerobe Agar (FAA) (Lab M)	46 g
Vancomycin (7.5 mg/ml)	1 ml
Neomycin (100 mg/ml)	1 ml
Distilled water	1000 ml
Sterile defibrinated sheep blood	50 ml
b. Preparation
   1) Combine all ingredients except sheep blood and vancomycin.
   2) Adjust pH to 7.0 to 7.2.
   3) Autoclave at 121°C for 15 minutes.
   4) Cool to 50°C, add blood and vancomycin, then pour plates.
c. Shelf life
   Plates - 10 days at refrigerator temperature
   Bottles - 1 year at refrigerator temperature

Hemin Solution (5 mg/ml)

a. Ingredients
Hemin (bovine)	0.5 g
Commercial ammonia water or sodium hydroxide (1N)	10 ml
Distilled water	90 ml

b. Preparation
   1) Dissolve hemin in ammonia water or sodium hydroxide (1N). Bring volume to
      100 ml with water.
   2) Autoclave at 121°C for 15 minutes.
c. Use
   Add to medium as a supplement in a final concentration of 5 μg/ml.

Kanamycin Solution (100 mg/ml)

a. Ingredients
Kanamycin	1 g
Sterile phosphate buffer, pH 8	10 ml
b. Preparation
   1) Dissolve kanamycin in phosphate buffer.
   2) Store in refrigerator for up to 1 year.

Kanamycin-Vancomycin Laked Blood Agar (KVLB)

a. Ingredients
Brucella agar (BBL or Difco)	43 g
Hemin solution (5 mg/ml)	1 ml
Vitamin K_1 solution (1 mg/ml)	1 ml
Kanamycin solution (100 mg/ml)	0.75 ml
Distilled water	1000 ml
Vancomycin solution (7.5 mg/ml or 2 mg/ml [KVLB-2])	1 ml
Laked sheep blood (blood frozen overnight, then thawed)	50 ml
b. Preparation
   1) Combine all ingredients except vancomycin and laked sheep blood.
   2) Boil to dissolve.
   3) Autoclave at 121°C for 15 minutes.
   4) Cool to 50°C, add vancomycin and laked sheep blood, then pour plates.
c. Shelf life
   Plates - 2 weeks at refrigerator temperature
   Bottles - 1 year at refrigerator temperature

Lactobacillus Selective Medium (LBS)

a. Ingredients
LBS agar (Rogosa-BBL)	8.4 g
Tomato juice (grocery store)	40 ml
Distilled water	60 ml
Acetic acid (glacial)	0.13 ml
b. Preparation
   1) Mix ingredients
   2) Steam for 20 to 30 minutes until agar dissolves or heat with frequent agitation
      and boil for 1 minute. Do not autoclave.
   3) Cool to 45°C and pour plates.
c. Shelf life
   Plates - 24 hours at room temperature

Freeze blood in glass or plastic containers filled $\leq\frac{3}{4}$ full (to avoid cracking) for a minimum of overnight. Thaw when ready to use.

Magnesium sulfate (10%)

a. Ingredients
    Magnesium sulfate   5 g
    Sterile distilled water   50 ml
b. Preparation
    1) Dissolve the magnesium sulfate in water.
    2) Sterilize by filtration or autoclave at 121 °C for 15 min.
    3) Store in refrigerator for up to 6 months.
c. Use
    Add 0.25 ml of magnesium sulfate along with 1% pyruvate into 10 ml of medium to stimulate the growth of sulfate-reducing organisms.

McFarland standards

a. Ingredients
    Barium chloride, 1.175% aqueous ($BaCl_2.2H_2O$; i.e., 0.048 M $BaCl_2$ Sulfuric acid, 1% aqueous (0.36 M)
b. Preparation
    1) Add amounts of the two solutions indicated in Table C-2 to tubes or ampules
    2) Seal
        Note: Tubes or ampules should have the same diameter as the tube to be used in subsequent density determinations.

0.5 McFarland standard

a. Ingredients
    Barium chloride, 1.175% aqueous   0.5 ml
    Sulfuric acid 1% aqueous   99.5 ml

Table C-2. Preparation of McFarland nephelometer standards.

Tube No.	1.175% Barium chloride (ml)	1% sulfuric acid (ml)	Corresponding approximate density of bacteria/ml
1	0.1	9.9	3×10^8
2	0.2	9.8	6×10^8
3	0.3	9.7	9×10^8
4	0.4	9.6	1.2×10^9
5	0.5	9.5	1.5×10^9
6	0.6	9.4	1.8×10^9
7	0.7	9.3	2.1×10^9
8	0.8	9.2	2.4×10^9
9	0.9	9.1	2.7×10^9
10	1.0	9.0	3×10^9

b. Preparation
    1) Mix the two solutions.
    2) Vortex and immediately dispense 4-6 ml into an appropriate tube or ampule and
        seal.
        **Note: the tube or ampule should have the same diameter as the tube to which
        it will be compared. Unless the standard is contained in a heat-sealed tube
        or ampule it should be replaced every 6 months. A 0.5 McFarland standard
        is assumed to be equivalent to approximately 1.5×10^8 organisms/ml.**
c. Store in the dark at room temperature.
d. Vigorously agitate the standard before each use.

Methanobacterium Medium (PRAS roll tubes)

Note: a gassing cannula apparatus must be used for roll tube preparation.

a. Ingredients for broth component

Rumen fluid (Animal Technologies, Inc.)	15 ml
Sodium chloride	0.2 g
Ammonium chloride	0.1 g
Monopotassium phosphate	0.1 g
Magnesium chloride, 6-hydrate	0.1 g
Calcium chloride	0.1 g
Ammonium molybdate, 4-hydrate	0.002 g
Cobalt chloride, 6-hydrate	0.002 g
Sodium bicarbonate	0.4 g
Resazurin solution	0.8 ml
L-cysteine hydrochloride	0.05 g
Distilled water	200 ml

b. Preparation
    1) Combine all ingredients except cysteine.
    2) Steam for 15 minutes (resazurin will stay pink).
    3) Cool while gassing with CO_2.
    4) Add cysteine, dissolve, adjust pH to 6.8. At this time, indicator will be color-
        less.
    5) Switch from CO_2 to 70% H_2 and 30% CO_2 (explosive mixture).
    6) Place 0.125 g agar (dry powder) in each roll tube.
    7) Dispense in 4.9 ml amounts, gassing with 70% H_2 and 30% CO_2.
    8) Autoclave tubes in a media press at 121°C for 15 minutes.
    9) Just before use, add 0.02 ml of sterile 5% Na_2S solution.
c. Shelf life - 3 months at room temperature.

Neomycin solution (100 mg/ml)

a. Ingredients

Neomycin	1 g
Sterile phosphate buffer, pH 8	10 ml

b. Preparation
    1) Dissolve neomycin in phosphate buffer.
    2) Store in refrigerator for up to 1 year.

Nessler reagent

a. Ingredients
Mercuric iodide	100 g
Potassium iodide	70 g
Sodium hydroxide (8N)	500 ml
Distilled water	~500 ml

b. Preparation
   1) Dissolve mercuric iodide and potassium iodide in 100 to 200 ml distilled water.
   2) Add this solution slowly and with stirring to the 8N sodium hydroxide at 20 to 25°C.
   3) Adjust volume to 1 liter with distilled water.
   4) Store in a rubber-stoppered bottle in the dark for up to 1 year.

Nitrate disks[257]

a. Ingredients
Potassium nitrate	30 g
Sodium molybdate dihydrate	0.1 g
Distilled water	100 ml
Sterile ¼ inch filter paper disks (BBL)	

b. Preparation
   1) Dissolve the nitrate and molybdate in the water.
   2) Sterilize by filtration.
   3) Dispense 20 μl quantities of the solution onto the disks arranged in a single layer in clean petri dishes.
   4) Allow the disks to dry at room temperature for 72 hours.
   5) Store at room temperature for 6 months.

Nitrate Reagents

a. Ingredients
   1) Solution A:
Sulfanilic acid	0.5 g
Glacial acetic acid	30 ml
Distilled water	120 ml
   2) Solution B:
1,6-Cleve's acid	0.2 g
Glacial acetic acid	30 g
Distilled water	120 ml

b. Preparation
   1) Dissolve ingredients of each solution in distilled water in separate containers.
   2) Store refrigerated for 3 months.

Oxgall (40%)

a. Ingredients
Oxgall	40 g
Distilled water	100 ml

b. Preparation
   1) Dissolve oxgall in water.
   2) Autoclave at 121°C for 15 minutes.
   3) Store refrigerated up to one year.

c. Use

Add 0.5 ml of oxgall solution to 10 ml of thioglycolate medium for bile inhibition test.

Paradimethylaminocinnamaldehyde [229]

a. Ingredients

Paradimethylaminocinnamaldehyde (Aldrich)	1 g
Hydrochloric acid (10%)	100 ml

b. Preparation

1) Dissolve paradimethylaminocinnamaldehyde in the hydrochloric acid solution.
2) Store in a dark bottle in refrigerator for two months.

c. Use for spot-indole test.

PMS (*Peptococcaceae* and *Megasphaera* Selective) Agar

a. Ingredients

Eggerth Gagnon (EG) agar (Eiken Chemical Co.)	40.7 g
PMS supplement solution (Eiken Chemical Co.)	10 ml
Distilled water	1000 ml

b. Preparation

1) Combine agar base and water.
2) Autoclave at 115°C for 20 minutes.
3) Cool to 50°C, add PS supplement solution and pour plates.

c. Shelf life unknown.

PS (*Peptostreptococcaceae* Selective) Agar

a. Ingredients

EG agar (Eiken Chemical Co.)	40.7 g
PS supplement solution (Eiken Chemical Co.)	10 ml
Phenylethyl alcohol	2 ml
Distilled water	1000 ml

b. Preparation

1) Combine agar base and water.
2) Autoclave at 115°C for 20 minutes.
3) Cool to 50°C, add PS supplement solution, phenylethyl alcohol, and pour plates.

c. Shelf life unknown.

Phenylethyl Alcohol Blood Agar (PEA)

a. Ingredients

Phenylethyl alcohol agar	42.5 g
Vitamin K_1 solution (10 mg/ml)	1 ml
Distilled water	1000 ml
Sterile defibrinated sheep blood	50 ml

b. Preparation

1) Combine ingredients except sheep blood.
2) Boil to dissolve.
3) Autoclave at 121°C for 15 minutes.
4) Cool to 50°C, add sheep blood, then pour plates.

c. Shelf life
    Plates - 2 weeks at refrigerator temperature
    Bottles - 1 year at refrigerator temperature

Pyruvate (20%)

a. Ingredients
    Pyruvic acid, sodium salt   20 g
    Distilled water   100 ml
b. Preparation
    1) Dissolve the pyruvate in water.
    2) Sterilize by filtration or autoclave at 121°C for 15 minutes.
    3) Store in refrigerator.

Resazurin

a. Ingredients
    Resazurin (Difco, Sigma)   1 tablet
    Distilled water   44 ml
b. Preparation
    1) Dissolve tablet in water.
    2) Store at room temperature.
c. Use for preparation of PRAS biochemicals

Rifampin (1 mg/ml)

a. Ingredients
    Rifampin   100 mg
    Absolute ethanol   20 ml
    Distilled water   80 ml
b. Preparation
    1) Dissolve rifampin in the alcohol.
    2) Add the distilled water.
    3) Store in refrigerator for up to 2 months.

Rifampin Blood Agar[231]

a. Ingredients
    Brucella agar (BBL or Difco)   43 g
    Hemin solution (5 mg/ml)   1 ml
    Vitamin K_1 solution (1 mg/ml)   1 ml
    Distilled water   1000 ml
    Sterile defibrinated sheep blood   50 ml
    Rifampin solution (1 mg/ml, see above)   50 ml
b. Preparation
    1) Combine all ingredients except sheep blood and rifampin.
    2) Boil to dissolve.
    3) Autoclave at 121°C for 15 minutes.
    4) Cool to 50°C, add rifampin and sheep blood, then pour plates.
c. Shelf life
    Plates - 1 week at refrigerator temperature
    Bottles - 1 year at refrigerator temperature

Ringer solution

a. Ingredients
Sodium chloride	9 g
Calcium chloride, dihydrate	0.25 g
Potassium chloride	0.4 g
Distilled water	1000 ml
b. Preparation
    1) Combine all ingredients.
    2) Mix to dissolve.
c. Use to prepare Ringer solution with metaphosphate, or Ringer dilution solution.

One-quarter strength Ringer solution with metaphosphate

a. Ingredients
Ringer solution	50 ml
Resazurin solution (see above)	0.8 ml
Distilled water	150 ml
Sodium metaphosphate	0.2 g
L-cysteine hydrochloride	0.1 g
b. Preparation
    1) Combine all ingredients.
    2) Mix to dissolve, dispense.
    3) Autoclave at 121°C for 15 minutes.

Ringer dilution solution

a. Ingredients
Ringer solution	200 ml
Resazurin solution	0.8 ml
L-cysteine hydrochloride	0.1 g
b. Preparation
    1) Combine all ingredients.
    2) Mix to dissolve, dispense 9 ml into disposable Hungate anaerobic culture tubes (Bellco).
    3) Autoclave at 121°C for 15 minutes.

Skim milk, 20%

a. Ingredients
Skim milk powder (BBL)	200 g
Distilled H_2O	1000 ml
b. Preparation
    1) Combine ingredients
    2) Dispense 0.25 - 0.5 ml into ½ dram vial.
    3) Autoclave at 110°C for 10 min with quick exhaust.
    4) Store refrigerated for up to 6 months.

Sodium bicarbonate (20 mg/ml)

a. Ingredients
Sodium bicarbonate	2 g
Distilled water	100 ml

b. Preparation
   1) Dissolve sodium bicarbonate in water.
   2) Sterilize by filtration.
   3) Store refrigerated for up to 6 months.
c. Use
   Add 0.5 ml of sodium bicarbonate solution to 10 ml of medium.

Sodium polyanethol sulfonate (SPS) disks [258]

a. Ingredients
Sodium polyanethol sulfonate (Sigma)	5 g
Distilled water	100 ml
Sterile ¼ inch filter paper disks (BBL)	
b. Preparation
   1) Dissolve sodium polyanethol sulfonate in water.
   2) Sterilize by filtration.
   3) Dispense 20 μl onto each filter paper disk as for preparation of nitrate disks.
   4) Allow to dry at room temperature for 72 hours.
   5) Store at room temperature for up to 6 months. (Prepared disks are available from Anaerobe Systems, Remel and Difco)

Thioglycolate, supplemented

a. Ingredients
Thioglycolate medium without indicator (BBL - 135C)	30 g
Hemin solution (5 mg/ml)	1 ml
Vitamin K_1 (1 mg/ml)	0.1 ml
Distilled water	1000 ml
b. Preparation
   1) Combine ingredients.
   2) Boil to dissolve.
   3) Dispense into tubes containing a marble chip (Fisher), filling the tubes two-thirds to three-fourths full.
   4) Autoclave at 118°C to 121°C for 15 minutes.
c. Use
   1) Just prior to use boil or steam for 5 minutes, cool.
   2) Supplement with normal rabbit or horse serum (10%) or Fildes enrichment (5%) if needed.
d. Shelf life - 6 months at room temperature

Tryptic Soy-Serum-Bacitracin-Vancomycin (TSBV) Agar

a. Ingredients
Trypticase soy agar (BBL)	40 g
Yeast extract	1 g
Distilled water	1000 ml
Sterile horse serum	100 ml
Bacitracin (7.5 mg/ml)	1 ml
Vancomycin (5 mg/ml)	1 ml
Vitamin K_1 (1 mg/ml)	1 ml
Hemin (5 mg/ml)	1 ml

b. Preparation
   1) Combine agar, yeast extract, and water.
   2) Adjust pH to 7.2.
   3) Boil to dissolve.
   4) Autoclave at 121°C for 15 minutes.
   5) Cool to 56°C, add sterile horse serum, bacitracin, and vancomycin, and pour plates.
c. Shelf life unknown

Tween-80

a. Ingredients

Tween-80 (Polysorbate 80)	10 ml
Distilled water	90 ml

b. Preparation
   1) Mix ingredients as much as possible.
   2) Autoclave at 121°C for 15 minutes and mix thoroughly while warm.
   3) Store refrigerated for up to 6 months.
c. Use
   Add 0.5 ml of Tween solution to 10 ml of medium to enhance growth of most gram-positive bacteria.

Urea broth

a. Ingredients

Urea broth powder (Difco)	38.7 g
Distilled water	1000 ml

b. Preparation
   1) Combine ingredients.
   2) Sterilize by filtration.
   3) Dispense into sterile tubes.
   4) Alternative method: Dilute commercial 10x urea broth (Difco or BBL) 1:10 in distilled water.
   5) Store refrigerated for up to 6 months.

Vancomycin stock solution (7.5 mg/ml)

a. Ingredients

Vancomycin	75 mg
Hydrochloric acid (0.05N)	5 ml
Distilled water	5 ml

b. Preparation
   1) Dissolve vancomycin in hydrochloric acid.
      Note: to make 4 mg/ml or 5 mg/ml stock solutions (for JVN and TSBV agars), use 40 mg or 50 mg of vancomycin, respectively.
   2) Add distilled water.
   3) Store in refrigerator for up to 1 month or in freezer (-20°C) for up to 1 year.

Veillonella-Neomycin Agar (VNA)

a. Ingredients

Veillonella agar (Difco)	36 g
Neomycin solution (100 mg/ml)	1 ml
Distilled water	1000 ml
Vancomycin solution (7.5 mg/ml)	1 ml

b. Preparation
    1) Combine all ingredients except vancomycin.
    2) Boil to dissolve.
    3) Autoclave at 121°C for 15 minutes.
    4) Cool to 50°C, add vancomycin and pour plates.
c. Shelf life
    Plates - 2 weeks at refrigerator temperature
    Bottles - 1 year at refrigerator temperature

Wilkins - Chalgren Agar

a. Ingredients

Trypticase	10 g
Gelysate (BBL)	10 g
Yeast extract	5 g
Dextrose	1 g
Sodium chloride	5 g
L-arginine, free base (Sigma)	1 g
Sodium pyruvate	1 g
Vitamin K_1	0.5 mg
Hemin	5 mg
Agar	5 mg
Distilled water	1000 ml

b. Preparation
    1) Combine ingredients
    2) Boil to dissolve.
    3) Dispense into screw cap tubes.
    4) Autoclave at 121°C for 15 minutes.
c. Shelf life - 4 weeks at refrigerator temperature

Vitamin K_1 solution (1 mg/ml)

a. Ingredients

Vitamin K_1 (Sigma)	0.1 g
Absolute ethanol	100 ml

b. Preparation
    1) Weigh out the vitamin K_1.
    2) Add this to a tube or bottle containing the ethanol.
    3) Store in refrigerator in tightly closed container, protected from light.
    4) Dilutions of this solution may be made in sterile distilled water.
c. Use
    To supplement any medium, add to a final concentration of 1 μl/ml.

VPI salts solution [109]

a. Ingredients

Calcium chloride (anhydrous)	0.2 g
Magnesium sulfate (anhydrous)	0.2 g
Dipotassium phosphate	1 g
Monopotassium phosphate	1 g
Sodium bicarbonate	10 g

Sodium chloride   2 g
Distilled water   1000 ml

b. Preparation
    1) Mix calcium chloride and magnesium sulfate in 300 ml water until dissolved.
    2) Add 500 ml water and while stirring add remaining salts.
    3) Continue stirring until all salts are dissolved.
    4) Add 200 ml water, mix.
    5) Store in refrigerator.

Appendix D

Stocking and Shipping Cultures

STOCKING CULTURES

Stock cultures should be prepared as soon as an organism is isolated in pure culture. Isolates can be put into stock from liquid or solid media. In either case, a young, actively growing culture should be used.

Supplemented thioglycolate medium or chopped meat broth incubated for 24 to 48 hours, depending upon the growth rate of the isolate, can be used to prepare stock cultures. Add 0.5 ml of the broth culture to an equal volume of sterile 20% skim milk prepared in a screw-capped ½-dram vial. Freeze and maintain the stock culture at -70°C. Plate out a portion of the broth culture to check the purity of the isolate.

If stock cultures are prepared from solid media, the growth taken from the plate or slant must be suspended carefully and mixed thoroughly in the skim milk. We have found that sterile wooden sticks work very well for this purpose.

Lyophilization (or freeze-drying) is an excellent method for preservation of stock cultures. Bacterial suspensions in a liquid medium (cooked meat broth) can be directly lyophilized; alternatively, a healthy culture grown on solid medium can be suspended in 20% skim milk and subsequently lyophilized. Both techniques allow for adequate recovery of the organism.

SHIPPING CULTURES

To ensure adequate recovery of the isolate, a fresh culture must be used to inoculate the transport system. A stoppered tube with an agar slant (chopped meat glucose) injected with a liquid culture suspension is simple and eliminates the need for paraffin-sealed tubes. If the technique is available, lyophilized preparations are preferred for their ease of maintenance and transportation. Cultures may also be injected into specimen transport vials (Port-a-Cul, B-D Microbiology Systems; Anaerobe Systems tube). Alternatively, grow the organism on an agar medium and use a swab to collect the growth and inoculate PRAS semi-solid transport medium (Anaerobe Systems).

Specimen mailer systems for handling diagnostic materials, complete with protective mailer, absorbent material, waterproof tape, labels, and shipping carton are commercially available (Polyfoam Packers Corp., O. Berk International, Inc.). Verify that the system conforms to Federal and postal regulations.

Regulations concerning packaging and shipment of etiologic agents are detailed in the Code of Federal Regulations, Section 72.25 of Part 72, Title 42, amended.

Sources of Supplies

Media and Reagents

AB Biodisk North American, Inc.
11260 Overland Ave. Suite 6A
Culver City, CA 90230

Adams Scientific, Inc.
771 Main St.
W. Warwick, RI 02893

Aldrich Chemical Co. Inc.
P.O. Box 355
Milwaukee, WI 53201

American Micro Scan
(Baxter Diagnostics, Inc.)
1584 Enterprise Blvd.
W. Sacramento, CA 95691

Anaerobe Systems
2200 C Zanker Rd.
San Jose, CA 95131

Analytab Products
Div. of Sherwood Medical
200 Express St.
Plainview, NY 11803

API (Bio Merieux SA)
69280 Marcy-l'Etoile
France

Animal Technologies, Inc.
P. O. Box 130243
Tyler, TX 75713

Baxter Healthcare Corp.,
Bartels Diagnostics Division
2005 NW Sammamish Rd., Ste. 107
Issaquah, WA 98027

Becton Dickinson Microbiology
Systems
P.O. Box 243
Cockeysville, MD 21030

Carr Scarborough
Microbiologicals, Inc.
P.O. Box 1328
Stone Mountain, GA 30083

Difco Laboratories
P.O. Box 1058A
Detroit, MI 48232

Eastman Kodak Co.
343 State St.
Rochester, NY 14652-3512

Eiken Chemical Co., LTD
33-8 Hongo 1-chome, Bunkyo-ku
Tokyo, Japan

Innovative Diagnostic Systems, Inc.
3404 Oakcliff Road, Ste. C-1
Atlanta, GA 30340

Lab M Ltd.
Topley House
POB 19, Bury,
Lancashire, BL9 6AU
United Kingdom

Mallinckrodt Science Products
675 McDonnell Blvd.
St. Louis, MO 63134

Merck, E.
Frankfurter Strasse 250, D-6100
Darmstadt 1,
West Germany

Meridian Diagnostics, Inc.
3471 River Hills Dr.
Cincinnati, OH 45244

MicroTech Medical Systems, Inc.
401 Laredo St.
Aurora, CO 80011

MSI/MicroMedia Systems
2330 Denison Ave.
Cleveland, OH 44109

Nissui Pharmaceutical Co., LTD.
2-11-1 Sugamo, Toshima-ku
Tokyo 170, Japan

Oxoid USA, Inc.
9017 Red Branch Rd.
Columbia, MD 20145

Pittman and Moore, Inc.
421 East Hawley
Mundelein, IL 60060

PML Microbiologicals
P.O. Box 578
Tualatin, Oregon 97062

Radiometer America (Sensititre)
811 Sharon Dr.
Westlake, OH 44145

Remel
12076 Sante Fe Dr.
Lenexa, KS 66215

Rosco Diagnostica
Taastrup, Denmark

Sigma Chemical Company
P. O. Box 14508
St. Louis, MO 63178

Vitek Systems
595 Anglum Dr.
Hazelwood, MO 63042

Equipment and supplies

Bellco Glass, Inc.
P.O. Box B
340 Edrudo Rd.
Vineland, NJ 08360

Coy Laboratory Products, Inc.
22 Metty Dr.
Ann Arbor, MI 48103

Craft Machine, Inc.
I-95 and Concord Rd.
Chester, PA 19013

Engelhard Industries
Gas Equipment Division
E. Newark, NJ 07029

Forma Scientific, Inc.
P.O. Box 649
Millcreek Rd.
Marietta, OH 45750

Hewlett-Packard Co.
Analytical Products Group
3000 Hanover St.
Palo Alto, CA 94304

MART BV Microbiology Automation
James Watt straat 11,
Postbus 165
7130 AD Lictenvoorde
Holland

Matheson Gas Products
30 Seaview Dr., P.O Box 1587
Secaucus, NJ 07096

Microbial ID, Inc.
115 Barksdale Professional Center
Newark, DE 19711

O. Berk International Inc.
3 Milltown Court
Union, New Jersey 07083

Packard Instrument Co.
1 State St.
Meriden, CT 06450

Polyfoam Packers Corp.
2320 S. Foster
Wheeling, IL 60090

Scientific Device Laboratory, Inc.
P.O. Box 88
Glenview, IL 60025

Sheldon Manufacturing Inc.
P.O. Box 627
300 N. 26th Ave.
Cornelius, OR 97113

Spectra-Physics
333 N. First St.
San Jose, CA 95134

Spiral System Instruments
7830 Old Georgetown Road
Bethesda, MD 20814

Tekmar Co.
7143 Kemper Rd.
P.O. Box 429576
Cincinnati, OH 45222-1856

Unimar, Inc.
475 Danbury Rd.
Wilton, CT 06897

Varian Associates, Inc.
220 Humboldt Ct.
3000 Hanover St.
Sunnyvale, CA 94089

Yamanouchi Pharmaceuticals
#5, 2-Chome,
Nihon bashi-Hancho, Chuo-ko
Tokyo, Japan

Reference cultures

American Type Culture Collection
12301 Parklawn Dr.
Rockville, MD 20852

National Collection of Type Cultures
Central Public Health Laboratory
Colindale Avenue
London NW9 5HT, England

References

1. **Adriaans, B. and H. Shah.** 1988. *Fusobacterium ulcerans* sp.nov.from tropical ulcers. Int. J. Syst. Bacteriol. **38:**447-448.

2. **Altemeier, W.A.** 1940. The anaerobic streptococci in tubo-ovarian abscess. Am. J. Obstet. Gynecol. **39:**1038-1042.

3. **Altemeier, W.A. and W.D. Fullen.** 1971. Prevention and treatment of gas gangrene. JAMA **217:**806-813.

4. **Appelbaum, P.C., C.S. Kaufman, J.C. Keifer and H.J. Venbrux.** 1984. Comparison of three methods for anaerobe identification. J. Clin. Microbiol. **18:**614-621.

5. **Aranki, A., S.A. Syed, E.B. Kenney and R. Freter.** 1969. Isolation of anaerobic bacteria from human gingiva and mouse cecum by means of a simplified glove-box procedure. Appl. Microbiol. **17:**568-576.

6. **Attebery, H.R. and W.T. Carter.** 1972. Rotator-pipette method for surface inoculation of petri plates. Abstr. Annu. Meeting, Am. Soc. Microbiol. p. 11.

7. **Ayyagari, A., V.K. Pancholi, S.C. Pandhi, A. Goswami, K.C. Agarwal and Y.N. Mehra.** 1981. Anaerobic bacteria in chronic suppurative otitis media. Indian J. Med. Res. **73:**860-864.

8. **Bailey, G.D., L.V.H. Moore, D.N. Love and J.L. Johnson.** 1988. *Bacteroides heparinolyticus:* Deoxyribonucleic acid relatedness of strains from the oral cavity and oral-associated disease conditions of horses, cats, and humans. Int. J. Syst. Bacteriol. **38:**42-44.

9. **Balows, A., W. J. Hausler, Jr., E. L. Herrmann, H. D. Isenberg and H. J. Shadomy.** 1991. Manual of Clinical Microbiology, 5th ed. ASM, Washington, D.C..

10. **Baron, E. J. and S. M. Finegold.** 1990. Bailey & Scott's Diagnostic Microbiology, 8th ed. C.V. Mosby Co., St. Louis.

11. **Baron, E.J., P. Summanen, J. Downes, M.C. Roberts and S.M. Finegold.** 1989. *Bilophila wadsworthia,* a unique gram-negative anaerobic rod recovered from appendicitis specimens and human feces. J. Gen. Microbiol. **135:**3405-3411.

12. **Baron, E.J., P. Summanen, J. Downes, M.C. Roberts, H. Wexler and S.M. Finegold.** 1990. Validation of the publication of new names and new combinations previously effectively published outside the IJSB. List No. 34. Int. J. Syst. Bacteriol. **40:**320-321.

13. **Barry, A.L., P.C. Fuchs, E.H. Gerlach, S.D. Allen, J.F. Acar, K.E. Aldridge, A.-M. Bourgault, H. Grimm, G.S. Hall, W. Heizmann, R.N. Jones, J.M. Swenson, C. Thornsberry, H. Wexler, J.D. Williams and J. Wüst.** 1990. Multi-laboratory evaluation of an agar diffusion disk susceptibility test for rapidly growing anaerobic bacteria. Rev. Infect. Dis. **12 Suppl. 2:**S210-S217.

14. **Bartlett, J.G. and S.L. Gorbach.** 1976. Anaerobic infections of the head and neck. Otolaryngol. Clin. North Am. **9:**655-678.

15. **Bartlett, J.G., S.L. Gorbach and S.M. Finegold.** 1974. The bacteriology of aspiration pneumonia. Am. J. Med. **56:**202-207.

16. **Bartlett, J.G., S.L. Gorbach, F.P. Tally and S.M. Finegold.** 1974. Bacteriology and treatment of primary lung abscess. Am. Rev. Respir. Dis. **109:**510-518.

17. **Bartlett, J.G., S.L. Gorbach, H. Thadepalli and S.M. Finegold.** 1974. The bacteriology of empyema. Lancet **1:**338-340.

18. **Bartlett, J.G., A.B. Onderdonk, E. Drude, C. Goldstein, M. Anderka, S. Alpert and W.M. McCormick.** 1977. Quantitative bacteriology of the vaginal flora. J. Infect. Dis. **136:**271-277.

19. **Bartlett, J.G. and B.F. Polk.** 1984. Bacterial flora of the vagina: Quantitative study. Rev. Infect. Dis. **6:**S67-S72.

20. **Beaucage, C.M. and A.B. Onderdonk.** 1982. Evaluation of prereduced anaerobically sterilized medium (PRAS II) system for identification of anaerobic microorganisms. J. Clin. Microbiol. **16:**570-572.

21. **Becker, G.D., G.J. Parell, D.F. Busch, S.M. Finegold and M.J. Acquarelli.** 1978. Anaerobic and aerobic bacteriology in head and neck cancer surgery. Arch. Otolaryngol. **104:**591-594.

22. **Beerens, H. and M. Tahon-Castel.** 1965. Infections Humaines à Bactéries Anaérobies Non-Toxigènes, Presses Acad.Eur., Bruxelles.

23. **Bennion, R.S., J.E. Thompson, E.J. Baron and S.M. Finegold.** 1990. Gangrenous and perforated appendicitis with peritonitis—treatment and bacteriology. Clin. Ther. **12:**31-44.

24. **Bjerkestrand, G., A. Digranes and A. Schreiner.** 1975. Bacteriological findings in transtracheal aspirates from patients with chronic bronchitis and bronchiectasis: a preliminary report. Scand. J. Respir. Dis. **56:**201-207.

25. **Brazier, J.S.** 1986. Yellow fluorescence of fusobacteria. Lett. Appl. Microbiol. **2:**125-126.

26. **Brazier, J.S., D.M. Citron and E.J.C. Goldstein.** 1991. A selective medium for *Fusobacterium* spp. J. Appl. Bacteriol. **71:**343-346

27. **Brazier, J.S., E.J.C. Goldstein, D.M. Citron and M.I. Ostovari.** 1990. Fastidious anaerobe agar compared with Wilkins-Chalgren agar, brain heart infusion agar, and brucella agar for susceptibility testing of *Fusobacterium* species. Antimicrob. Agents Chemother. **34:**2280-2282.

28. **Brazier, J.S. and T.V. Riley.** 1988. UV red fluorescence of *Veillonella* spp. J. Clin. Microbiol. **26:**383-384.

29. **Brazier, J.S. and S.A. Smith.** 1989. Evaluation of the Anoxomat: a new technique for anaerobic and microaerophilic clinical bacteriology. J. Clin. Pathol. **42:**640-644.

30. **Brondz, I. and J. Carlsson.** 1990. Cellular fatty acids of *Bacteroides forsythus*. J. Chromatogr. Biomed. Appl. **533:**141-144.

31. **Brook, I.** 1987. Comparison of two transport systems for recovery of aerobic and anaerobic bacteria from abscesses. J. Clin. Microbiol. **25:**2020-2022.

32. **Brook, I.** 1988. Microbiology of non-puerperal breast abscesses. J. Infect. Dis. **157:**377-379.

33. **Brook, I. and S.M. Finegold.** 1979. Bacteriology of chronic otitis media. JAMA **241:**487-488.

34. **Brook, I., E.H. Frazier and M.E. Gher.** 1991. Aerobic and anaerobic microbiology of periapical abscess. Oral Microbiol. Immunol. **6:**123-125.

35. **Brook, I., E.H. Frazier and D.H. Thompson.** 1991. Aerobic and anaerobic microbiology of peritonsillar abscess. Laryngoscope **101:**289-292.

36. **Brook, I., S. Grimm and R.B. Kielich.** 1981. Bacteriology of acute periapical abscess in children. J. Endod. **7:**378-380.

37. **Broughton, W.A., J.B. Bass and M.B. Kirkpatrick.** 1987. The technique of protected brush catheter bronchoscopy. J. Crit. Ill. **2:**63-70.

38. **Buchanan, A.G.** 1982. Clinical laboratory evaluation of a reverse CAMP test for presumptive identification of *Clostridium perfringens*. J. Clin. Microbiol. **16:**761-762.

39. **Buchanan, R. E. and N. E. Gibbons (eds.).** 1974. Bergey's Manual of Determinative Bacteriology, 8th ed. Williams & Wilkins, Baltimore.

40. **Burlage, R.S. and P.D. Ellner.** 1985. Comparison of the PRAS II, AN-Ident, and RapID-ANA systems for identification of anaerobic bacteria. J. Clin. Microbiol. **22:**32-35.

41. **Cato, E.P., L.V.H. Moore and W.E.C. Moore.** 1985. *Fusobacterium alocis* sp.nov.and *Fusobacterium sulci* sp.nov.from the human gingival sulcus. Int. J. Syst. Bacteriol. **35:**475-477.

42. **Charfreitag, O., M.D. Collins and E. Stackebrandt.** 1988. Reclassification of *Arachnia propionica* as *Propionibacterium propionicus* comb.nov. Int. J. Syst. Bacteriol. **38:**354-357.

43. **Chastre, J., J.Y. Fagon, P. Soler, M. Bornet, Y. Domart, J.L. Trouillet, C. Gibert and A.J. Hance.** 1988. Diagnosis of nosocomial bacterial pneumonia in intubated patients undergoing ventilation: comparison of the usefulness of bronchoalveolar lavage and the protected specimen brush [published erratum appears in Am J Med 1989 Feb;86(2):258]. Am. J. Med. **85:**499-506.

44. **Chow, A.W., J.E. Galpin and L.B. Guze.** 1977. Clindamycin for treatment of sepsis caused by decubitus ulcers. J. Infect. Dis. **135(Suppl.):**S65-S68.

45. **Chow, A.W., K.L. Malkasian, J.R. Marshall and L.B. Guze.** 1975. The bacteriology of acute pelvic inflammatory disease. Am. J. Obstet. Gynecol. **122:**876-879.

46. **Chow, A.W., J.R. Marshall and L.B. Guze.** 1977. A double-blind comparison of clindamycin with penicillin plus chloramphenicol in treatment of septic abortion. J. Infect. Dis. **135:**S35-S39.

47. **Chow, A.W., S.M. Roser and F.A. Brady.** 1978. Orofacial odontogenic infections. Ann. Intern. Med. **88:**392-402.

48. **Citron, D.M., E.J. Baron, S.M. Finegold and E.J.C. Goldstein.** 1990. Short prereduced anaerobically sterilized (PRAS) biochemical scheme for identification of clinical isolates of bile-resistant *Bacteroides* species. J. Clin. Microbiol. **28:**2220-2223.

49. **Citron, D.M., M.I. Ostovari, A. Karlsson and E.J.C. Goldstein.** 1991. Evaluation of the E test for susceptibility testing of anaerobic bacteria. J. Clin. Microbiol. **29:**2197-2203.

50. **Collins, M.D. and H.N. Shah.** 1986. Reclassification of *Bacteroides termitidis* Sebald (Holdeman and Moore) in a new genus *Sebaldella,* as *Sebaldella termitidis* comb.nov. Int. J. Syst. Bacteriol. **36:**349-350.

51. **Collins, M.D. and H.N. Shah.** 1986. Reclassification of *Bacteroides praeacutus* Tissier (Holdeman and Moore) in a new genus, *Tissierella,* as *Tissierella praeacuta* comb.nov. Int. J. Syst. Bacteriol. **36:**461-463.

52. **Collins, M.D., H.N. Shah and T. Mitsuoka.** 1985. Reclassification of *Bacteroides microfusus* (Kaneuchi and Mitsuoka) in a new genus *Rikenella,* as *Rikenella microfusus* comb.nov. Syst. Appl. Microbiol. **6:**79-81.

53. **Collins, M.D., H.N. Shah and T. Mitsuoka.** 1985. Validation of the publication of new names and new combinations previously effectively published outside the IJSB. List No.18. Int. J. Syst. Bacteriol. **35:**375-376.

54. **Cooper, S.W., D.G. Pfeiffer and F.P. Tally.** 1985. Evaluation of xylan fermentation for the identification of *Bacteroides ovatus* and *Bacteroides thetaiotaomicron.* J. Clin. Microbiol. **22:**125-126.

55. **Crawford, J.J., J.R. Sconyers and J.D. Moriarty.** 1974. Bacteremia after tooth extraction studied with the aid of pre-reduced anaerobically sterilized culture media. Appl. Microbiol. **27:**927.

56. **Dellinger, C.A. and L.V.H. Moore.** 1986. Use of the RapID-ANA System to screen for enzyme activities that differ among species of bile-inhibited Bacteroides. J. Clin. Microbiol. **23:**289-293.

57. **Dewhirst, F.E., B.J. Paster, S. La Fontaine and J.I. Rood.** 1990. Transfer of *Kingella indologenes* (Snell and Lapage 1976) to the genus *Suttonella* gen.nov.as *Suttonella indologenes* comb.nov.; transfer of *Bacteroides nodosus* (Beveridge 1941) to the genus *Dichelobacter* gen.nov.as *Dichelobacter nodosus* comb.nov.; and assignment of the genera *Cardiobacterium, Dichelobacter,* and *Suttonella* to *Cardiobacteriaceae* fam.nov.in the gamma division of *Proteobacteria* on the basis of *16S rRNA* sequence comparisons. Int. J. Syst. Bacteriol. **40:**426-433.

58. **Dowell, V. R., Jr. and T. M. Hawkins.** 1974. Laboratory Methods in Anaerobic Bacteriology (Spanish edition also), US Govt, Washington, D.C..

59. **Dowell, V. R., Jr. and G. L. Lombard.** 1981. Reactions of Anaerobic Bacteria in Differential Agar Media, U.S. Department of Health and Human Services, Public Health Service, Centers for Disease Control, Atlanta.

60. **Dowell, V. R., Jr., G. L. Lombard, F. S. Thompson and A. Y. Armfield.** 1977. Media for Isolation, Characterization, and Identification of Obligately Anaerobic Bacteria, Centers for Disease Control, Atlanta.

61. **Downes, J., J.I. Mangels, J. Holden, M.J. Ferraro and E.J. Baron.** 1990. Evaluation of two single-plate incubation systems and the anaerobic chamber for the cultivation of anaerobic bacteria. J. Clin. Microbiol. **28:**246-248.

62. **Draper, D.L. and A.L. Barry.** 1977. Rapid identification of *Bacteroides fragilis* with bile and antibiotic disks. J. Clin. Microbiol. **5:**439-443.

63. **Drasar, D. R. and M. J. Hill.** 1975. Human Intestinal Flora, Academic Press, London.

64. **Driscoll, C.** 1990. Minimally invasive endometrial sampling. Patient Care Feb.15: 206-208.

65. **Edelstein, M.A.C., D.M. Citron and M.E. Mulligan.** 1983. Preparation and storage of selective media for *Clostridium difficile,* Abstr.Annu.Meeting (Abstract C-193). Abstr. Annu. Meeting, Am. Soc. Microbiol. p. 344.

66. **Edmiston, C.E., Jr., A.P. Walker, C.J. Krepel and C. Gohr.** 1990. The nonpuerperal breast infection: aerobic and anaerobic microbial recovery from acute and chronic disease. J. Infect. Dis. **162:**695-699.

67. **Ehrenkranz, N.J., B. Alfonso and D. Nerenberg.** 1990. Irrigation-aspiration for culturing draining decubitus ulcers: correlation of bacteriological findings with a clinical inflammatory scoring index. J. Clin. Microbiol. **28:**2389-2393.

68. **Embley, T.M., N. Faquir, W. Bossart and M.D. Collins.** 1989. *Lactobacillus vaginalis* sp.nov.from the human vagina. Int. J. Syst. Bacteriol. **39:**368-370.

69. **England, D.M. and J.E. Rosenblatt.** 1977. Anaerobes in human biliary tracts. J. Clin. Microbiol. **6:**494-498.

70. **Eschenbach, D.A., T.M. Buchanan, H.M. Pollock, P.S. Forsyth, E.R. Alexander, J.S. Lin, S.P. Wang, B.B. Wentworth, W.M. MacCormack and K.K. Holmes.** 1975. Polymicrobial etiology of acute pelvic inflammatory disease. N. Engl. J. Med. **293:**166-171.

71. **Eschenbach, D.A., P.R. Davick, B.L. Williams, S.J. Klebanoff, K. Young-Smith, C.M. Critchlow and K.K. Holmes.** 1989. Prevalence of hydrogen per-oxide-producing *Lactobacillus* species in normal women and women with bacterial vaginosis. J. Clin. Microbiol. **27:**251-256.

72. **Ezaki, T., S.-L. Liu, Y. Hashimoto and E. Yabuuchi.** 1990. *Peptostreptococcus hydrogenalis* sp.nov.from human fecal and vaginal flora. Int. J. Syst. Bacteriol. **40:**305-306.

73. **Farrow, J.A.E. and M.D. Collins.** 1988. *Lactobacillus oris* sp. nov. from the human oral cavity. Int. J. Syst. Bacteriol. **38:**116-118.

74. **Felner, J.M. and V.R. Dowell, Jr.** 1971. "*Bacteroides*" bacteremia. Am. J. Med. **50:**787-796.

75. **Finegold, S. M.** 1977. Anaerobic Bacteria in Human Disease, Academic Press, New York.

76. **Finegold, S.M.** 1989. Anaerobic pulmonary infection. Hosp. Pract. **24:**103-133.

77. **Finegold, S. M. and M. A. C. Edelstein.** 1988. Coping with anaerobes in the 80's. p. 1-10. In J. M. Hardie and S. P. Borriello (eds.), Anaerobes Today, John Wiley & Sons, Chichester, England.

78. **Finegold, S. M. and W. L. George (eds.).** 1989. Anaerobic Infections in Humans, Academic Press, San Diego,CA.

79. **Finegold, S. M., P. T. Sugihara and V. L. Sutter.** 1971. Use of selective media for isolation of anaerobes from humans, p. 99-108. In D. A. Shapton and R. G. Board (eds.), Isolation of Anaerobes. Soc.Appl.Bacteriol. Technical Series 5, Academic Press, London.

80. **Finegold, S. M., V. L. Sutter and G. E. Mathisen.** 1983. Normal indigenous intestinal flora, Chap.1, p. 3-31. In D. J. Hentges (ed.), Human Intestinal Micro-flora in Health and Disease, Academic Press, New York.

81. **Flora, D., P. Wideman, V.L. Sutter and S.M. Finegold.** 1975. Unpublished data.

82. **Flödstrom, A. and H.O. Hallander.** 1976. Microbiological aspects of peritonsillar abscesses. Scand. J. Infect. Dis. **8:**157-160.

83. **Frederick, J. and A.I. Braude.** 1974. Anaerobic infection of the paranasal sinuses. N. Engl. J. Med. **290:**135-137.

84. **Fukumoto, O.K. and Y. Takazoe.** 1988. Enumeration of cultivable black-pigmented *Bacteroides* species in human subgingival dental plaque and fecal samples. Oral Microbiol. Immunol. **3:**28-31.

85. **George, W.L., V.L. Sutter, D. Citron and S.M. Finegold.** 1979. Selective and differential medium for isolation of *Clostridium difficile*. J. Clin. Microbiol. **9:**214-219.

86. **Ghanem, F.M., A.C. Ridpath, W.E.C. Moore and L.V.H. Moore.** 1991. Identifi-cation of *Clostridium botulinum, Clostridium argentinense,* and related organisms by cellular fatty acid analysis. J. Clin. Microbiol. **29:**1114-1124.

87. **Gharbia, S.E. and H.N. Shah.** 1989. Glutamate dehyrogenase and 2-oxoglutarate reductase electrophoretic patterns and deoxyribonucleic acid-deoxyribonucleic acid hybridization among human oral isolates of *Fusobacterium nucleatum.* Int. J. Syst. Bacteriol. **39:**467-470.

88. **Gharbia, S.E. and H.N. Shah.** 1990. Heterogeneity within *Fusobacterium nucleatum,* proposal of four subspecies. Lett. Appl. Microbiol. **10:**105-108.

89. **Gmür, R.** 1988. Applicability of monoclonal antibodies to quantitatively monitor subgingival plaque for specific bacteria. Oral Microbiol. Immunol. **3:**187-191.

90. **Goldstein, E. J. C.** 1989. Bite infections, p. 455-465. In S. M. Finegold and W. L. George (eds.), Anaerobic Infections in Humans, Academic Press, San Diego.

91. **Gonzales-C, C.L. and F.M. Calia.** 1975. Bacteriologic flora of aspiration induced pulmonary infections. Ann. Intern. Med. **135:**711-714.

92. **Goodman, A.D.** 1977. Isolation of anaerobic bacteria from the root canal systems of necrotic teeth by use of a transport solution. Oral Surg. **43:**766-770.

93. **Gorbach, S.L.** 1975. Management of anaerobic infections: Intra-abdominal sepsis. Ann. Intern. Med. **83:**377-379.

94. **Gorbach, S.L., J.W. Mayhew, J.G. Bartlett, H. Thadepalli and A.B. Onderdonk.** 1976. Rapid diagnosis of anaerobic infections by direct gas-liquid chromatography of clinical specimens. J. Clin. Invest. **57:**478-484.

95. **Gorbach, S. L., H. Thadepalli and J. Norsen.** 1974. Anaerobic microorganisms in intraabdominal infections, p. 399-407. In A. Balows, R.M. DeHaan, V.R. Dowell, Jr. and L.B. Guze (eds.), Anaerobic Bacteria: Role in Disease, Charles C. Thomas, Springfield, IL.

96. **Greey, P.H.** 1932. The bacteriology of bronchiectasis. An analysis based on nine cases in which lobectomy was done. J. Infect. Dis. **50:**203-212.

97. **Gulletta, E., G. Amato, E. Nani and I. Covelli.** 1985. Comparison of two systems for identification of anaerobic bacteria. Eur. J. Clin. Microbiol. **4:**282-285.

98. **Haapasalo, M., H. Ranta, H. Shah, K. Ranta, K. Lounatmaa and R.M. Kroppenstedt.** 1986. *Mitsuokella dentalis* sp.nov.from dental root canals. Int. J. Syst. Bacteriol. **36:**556-568.

99. **Hall, W.L., A.I. Sobel, C.P. Jones and R.T. Parker.** 1967. Anaerobic postoperative pelvic infections. Obstet. Gynecol. 30:1-7.

100. **Hamory, B.H., M.A. Sande, A. Sydnor, Jr., D.L. Seale and J.M. Gwaltney, Jr.** 1979. Etiology and antimicrobial therapy of acute maxillary sinusitis. J. Infect. Dis. **139:**197-202.

101. **Han, Y.-H., R.M. Smibert and N.R. Krieg.** 1991. *Wolinella recta, Wolinella curva, Bacteroides ureolyticus,* and *Bacteroides gracilis* are microaerophiles, not anaerobes. Int. J. Syst. Bacteriol. **41:**218-222.

102. **Hansen, M.V. and L.P. Elliott.** 1980. New presumptive identification test for *Clostridium perfringens*: reverse CAMP test. J. Clin. Microbiol. **12:**617-619.

103. **Hansen, S.L. and B.J. Stewart.** 1976. Comparison of API and Minitek to Centers for Disease Control methods for the biochemical characterization of anaerobes. J. Clin. Microbiol. **4:**227-231.

104. **Head, C.B. and S. Ratnam.** 1988. Comparison of API ZYM system with API AN-Ident, API 20-A, Minitek Anaerobe II, and RapID-ANA systems for identification of *Clostridium difficile.* J. Clin. Microbiol. **26:**144-146.

105. **Heineman, H.S. and A.I. Braude.** 1963. Anaerobic infection of the brain. Am. J. Med. **35:**682-697.

106. **Helstad, A.G., M.A. Hutchinson, W.P. Amos and T.A. Kurzynski.** 1984. Evaluation of two broth disk methods for antibiotic susceptibility of anaerobes. Antimicrob. Agents Chemother. **26:**601-603.

107. **Hill, G.B. and S. Schalkowsky.** 1990. Development and evaluation of the spiral gradient endpoint method for susceptibility testing of anaerobic gram-negative bacilli. Rev. Infect. Dis. **12:**200-209.

108. **Hofstad, T.** 1980. Evaluation of the API ZYM system for identification of *Bacteroides* and *Fusobacterium* species. Med. Microbiol. Immunol. (Berl.) **168:**173-177.

109. **Holdeman, L.V., E.P. Cato and W.E.C. Moore.** 1977. Anaerobe Laboratory Manual, 4th ed. Virginia Polytechnic Institute and State University, Blacksburg, VA.

110. **Holdeman, L.V. and J.L. Johnson.** 1982. Descriptions of *Bacteroides loeschii* sp.nov. and emendation of the descriptions of *Bacteroides melaninogenicus* (Oliver and Wherry) Roy and Kelly 1939 and *Bacteroides denticola* Shah and Collins 1981. Int. J. Syst. Bacteriol. **32:**399-409.

111. **Holt, J.G. and N.R. Krieg.** 1984. Bergey's Manual of Systematic Bacteriology, Ninth Edition, Subvolume 1, The Williams and Wilkins Co., Baltimore, MD.

112. **Hovig, B. and E.H. Aandahl.** 1969. A selective method for isolation of *Haemophilus* in material from the respiratory tract. Acta Pathol. Microbiol. Scand. **77:**676-684.

113. **Hungate, R.E.** 1969. A roll tube method for cultivation of strict anaerobes, p. 117-132. In J.R. Norris and D.W. Robbons (eds.), Methods in Microbiology, Vol.3B, Academic Press, New York.

114. **Husain, M., K. Rajashekariah, K. Menda, J. Norsen and C. Kallick.** 1975. Anaerobic microbiology of soft tissue abscesses. Proc. 15th ICAAC Abstr. **56.**

115. **Hussain, Z., R. Lannigan, B.C. Schieven, L. Stoakes, T. Kelly and D. Groves.** 1987. Comparison of RapID ANA and Minitek with conventional method for biochemical identification of anaerobes. Diagn. Microbiol. Infect. Dis. **6:**69-72.

116. **Iino, Y., E. Hoshino, S. Tomioka, T. Takasaka, Y. Kaneko and R. Yuasa.** 1983. Organic acids and anaerobic microorganisms in the contents of the cholesteatoma sac. Ann. Otol. Rhinol. Laryngol. **92:**91-96.

117. **Iwen, P.C., S.J. Booth and G.L. Woods.** 1989. Comparison of media for screening of diarrheic stools for the recovery of *Clostridium difficile*. J. Clin. Microbiol. **27:**2105-2106.

118. **Iwu, C., T.W. MacFarlane, D. MacKenzie and D. Stenhouse.** 1990. The microbiology of periapical granulomas. Oral Surg. Oral Med. Oral Pathol. **69:**502-505.

119. **Jensen, N.S. and E. Canale-Parola.** 1986. *Bacteroides pectinophilus* sp. nov. and *Bacteroides galacturonicus* sp. nov.: Two pectinolytic bacteria from the human intestinal tract. Appl. Environ. Microbiol. **52:**880-887.

120. **Jensen, N.S. and E. Canale-Parola.** 1987. Validation of the publication of new names and new combinations previously effectively published outside the IJSB. List No.23. Int. J. Syst. Bacteriol. **37:**179-180.

121. **Jilly, B.J., P.C. Schreckenberger and L.J. LeBeau.** 1984. Rapid glutamic acid decarboxylase test for identification of *Bacteroides* and *Clostridium* spp. J. Clin. Microbiol. **19:**592-593.

122. **Johnson, C.C., J.F. Reinhardt, M.A.C. Edelstein, M.E. Mulligan, W.L. George and S.M. Finegold.** 1985. *Bacteroides gracilis,* an important anaerobic bacterial pathogen. J. Clin. Microbiol. **22:**799-802.

123. **Johnson, C.C., H.M. Wexler, S. Becker, M. Garcia and S.M. Finegold.** 1989. Cell-wall-defective variants of *Fusobacterium.* Antimicrob. Agents Chemother. **33:**369-372.

124. **Johnson, J.L., L.V.H. Moore, B. Kaneko and W.E.C. Moore.** 1990. *Actinomyces georgiae* sp.nov., *Actinomyces gerencseriae* sp.nov., designation of two genospecies of *Actinomyces naeslundii,* and *viscosus* serotype II in *A. naeslundii* genospecies 2. Int. J. Syst. Bacteriol. **40**:273-286.

125. **Johnson, J.L., W.E.C. Moore and L.V.H. Moore.** 1986. *Bacteroides caccae* sp.nov., *Bacteroides merdae* sp.nov., and *Bacteroides stercoris* sp.nov. isolated from human feces. Int. J. Syst. Bacteriol. **36**:499-501.

126. **Johnson, L.L., L.V. McFarland, P. Dearing, V. Raisys and F.D. Schoenknecht.** 1989. Identification of *Clostridium difficile* in stool specimens by culture-enhanced gas-liquid chromatography. J. Clin. Microbiol. **27**:2218-2221.

127. **Johnston, B.L., M.A.C. Edelstein, E.Y. Holloway and S.M. Finegold.** 1987. Bacteriologic and clinical study of *Bacteroides oris* and *Bacteroides buccae.* J. Clin. Microbiol. **25**:491-493.

128. **Jokipii, A.M., P. Karma, K. Ojala and L. Jokipii.** 1977. Anaerobic bacteria in chronic otitis media. Arch. Otolaryngol. **103**:278-280.

129. **Jones, D.B. and N.M. Robinson.** 1977. Anaerobic ocular infections. Trans. Am. Acad. Ophthalmol. Otolaryngol. **83**:309-331.

130. **Jousimies-Somer, H.R., S. Savolainen and J.S. Ylikoski.** 1988. Bacteriological findings of acute maxillary sinusitis in young adults. J. Clin. Microbiol. **26**:1919-1925.

131. **Jungkind, D., J. Millan, S. Allen, J. Dyke and E. Hill.** 1986. Clinical comparison of a new automated infrared blood culture system with the Bactec 460 system. J. Clin. Microbiol. **23**:262-266.

132. **Kaczmarek, F.S. and A.L. Coykendall.** 1980. Production of phenylacetic acid by strains of *Bacteroides assacharolyticus* and *Bacteroides gingivalis* (sp.nov.). J. Clin. Microbiol. **12**:288-290.

133. **Karachewski, N.O., E.L. Busch and C.L. Wells.** 1985. Comparison of PRAS II, RapID ANA, and API 20A systems for identification of anaerobic bacteria. J. Clin. Microbiol. **21**:122-126.

134. **Kato, N., C.-Y. Ou, H. Kato, S.L. Bartley, V.K. Brown, V.R. Dowell, Jr. and K. Ueno.** 1991. Identification of toxigenic *Clostridium difficile* by the polymerase chain reaction. J. Clin. Microbiol. **29**:33-37.

135. **Killgore, G.E., S.E. Starr, V.E. Del Bene, D.N. Whaley and V.R. Dowell, Jr.** 1973. Comparison of three anaerobic systems for the isolation of anaerobic bacteria from clinical specimens. Am. J. Clin. Pathol. **59**:552-559.

136. **Kilpper-Balz, R. and K.H. Schleifer.** 1988. Transfer of *Streptococcus morbillorum* to the genus *Gemella* as *Gemella morbillorum* comb.nov. Int. J. Syst. Bacteriol. **38**:442-443.

137. **Kitch, T.T. and P.C. Appelbaum.** 1989. Accuracy and reproducibility of the 4-hour ATB 32A method for anaerobe identification. J. Clin. Microbiol. **27**:2509-2513.

138. **Kodaka, H., A.Y. Armfield, G.L. Lombard and V.R. Dowell, Jr.** 1982. Practical procedure for demonstrating bacterial flagella. J. Clin. Microbiol. **16**:948-952.

139. **Koransky, J.R., S.D. Allen and V.R. Dowell, Jr.** 1978. Use of ethanol for selective isolation of sporeforming microorganisms. Appl. Environ. Microbiol. **35**:762-765.

140. **Kornman, K.S., M.G. Newman, R. Alvarado, T.F. Flemmig, S. Nachnani and J. Tumbusch.** 1991. Clinical and microbiological patterns of adults with peridontitis. J. Periodontol. **62**:634-642.

141. **Krohn, M.A., S.L. Hillier and D.A. Eschenbach.** 1989. Comparison of methods for diagnosing bacterial vaginosis among pregnant women. J. Clin. Microbiol. **27:**1266-1271.

142. **Lai, C.-H., B.M. Males, P.A. Dougherty, P. Berthold and M.A. Listgarten.** 1983. *Centipeda periodontii* gen.nov.,sp.nov.from human periodontal lesions. Int. J. Syst. Bacteriol. **33:**628-635.

143. **Laughon, B.E., S.A. Syed and W.J. Loesche.** 1982. Rapid identification of *Bacteroides gingivalis.* J. Clin. Microbiol. **15:**345-346.

144. **Laughon, B.E., S.A. Syed and W.J. Loesche.** 1982. API ZYM system for identification of *Bacteroides* spp., *Capnocytophaga* spp., and spirochetes of oral origin. J. Clin. Microbiol. **15:**97-102.

145. **Leach, R.D., S.J. Eykyn, I. Phillips and B. Corrin.** 1979. Anaerobic subareolar breast abscess. Lancet **1:**35-37.

146. **Ledger, W.J., C.L. Gee, R. Pollin, R.M. Nakamura and W.P. Lewis.** 1976. The use of pre-reduced media and a portable jar for the collection of anaerobic organisms from clinical sites of infection. Am. J. Obstet. Gynecol. **125:**677-681.

147. **Lee, K., E.J. Baron, P. Summanen and S.M. Finegold. 1990.** Selective medium for isolation of *Bacteroides gracilis.* J. Clin. Microbiol. **28:**1747-1750.

148. **Levison, M.E., I. Trestman, R. Quach, C. Sladowski and C.N. Floro.** 1979. Quantitative bacteriology of the vaginal flora in vaginitis. Am. J. Obstet. Gynecol. **133:**139-144.

149. **Lewis, R.P., V.L. Sutter and S.M. Finegold. 1978.** Bone infections involving anaerobic bacteria. Medicine(Baltimore) **57:**279-305.

150. **Livingston, S.J., S.D. Kominos and R.B. Yee.** 1978. New medium for selection and presumptive identification of the *Bacteroides fragilis* group. J. Clin. Microbiol. **7:**448-453.

151. **Loesche, W.J., S.A. Syed, E. Schmidt and E.C. Morrison.** 1985. Bacterial profiles of subgingival plaques in periodontitis. J. Periodontol. **56:**447-456.

152. **Lombard, G.L., D.N. Whaley and V.R. Dowell, Jr.** 1982. Comparison of media in the Anaerobe-Tek and Presumpto plate systems and evaluation of the Anaerobe-Tek system for identification of commonly encountered anaerobes. J. Clin. Microbiol. **16:**1066-1072.

153. **Lorber, B. and R.M. Swenson.** 1974. Bacteriology of aspiration pneumonia. A prospective study of community and hospital acquired cases. Ann. Intern. Med. **81:**329-331.

154. **Louie, T.J., J.G. Bartlett, F.P. Tally and S.L. Gorbach.** 1976. Aerobic and anaerobic bacteria in diabetic foot ulcers. Ann. Intern. Med. **85:**461-463.

155. **Love, D.N., J.L. Johnson, R.F. Jones, M. Bailey and A. Calverley.** 1986. *Bacteroides tectum* sp.nov.and characteristics of other nonpigmented *Bacteroides* isolates from soft-tissue infections from cats and dogs. Int. J. Syst. Bacteriol. **36:**123-128.

156. **Love, D.N., J.L. Johnson, R.F. Jones and A. Calverley.** 1987. *Bacteroides salivosus* sp.nov., an asaccharolytic, black-pigmented species from cats. Int. J. Syst. Bacteriol. **37:**307-309.

157. **Ludwig, W., M. Weizenegger, R. Kilpper-Balz and K.H. Schleifer.** 1988. Phylogenetic relationships of anaerobic streptococci. Int. J. Syst. Bacteriol. **38:**15-18.

158. **MacLennan, J.D.** 1962. The histotoxic clostridial infections of man. Bacteriol. Rev. **26:**177-276.

159. **Malnick, H., A. Jones and J.C. Vickers.** 1989. *Anaerobiospirillum*: Cause of a "new" zoonosis? [letter]. Lancet **1**:1145-1146.

160. **Malnick, H., M.E.M. Thomas, H. Lotay and M. Robbins.** 1983. *Anaerobiospirillum* species isolated from humans with diarrhoea. J. Clin. Pathol. **36**:1097-1101.

161. **Mardh, P.-A., L. Larsson and G. Odham.** 1981. Head-space gas chromatography as a tool in the identification of anaerobic bacteria and diagnosis of anaerobic infections. Scand. J. Infect. Dis. **Suppl.26**:14-18.

162. **Margaret, B.S. and G.N. Krywolap.** 1986. *Eubacterium yurii* subsp. *yurii* sp.nov. and *Eubacterium yurii* subsp. *margaretiae* subsp.nov.: Test tube brush bacteria from subgingival dental plaque. Int. J. Syst. Bacteriol. **36**:145-149.

163. **Marler, L., S. Allen and J. Siders.** 1984. Rapid enzymatic characterization of clinically encountered anaerobic bacteria with the API ZYM system. Eur. J. Clin. Microbiol. **3**:294-300.

164. **Martens, M.D. and et al.** 1989. Transcervical uterine cultures with a new endometrial suction curette: a comparison of three sampling methods in postpartum endometritis. Obstet. Gynecol. **74**:273-276.

165. **Masfari, A.M., B.I. Duerden and G.R. Kinghorn.** 1986. Quantitative studies of vaginal bacteria. Genitourin. Med. **62**:256-263.

166. **Mazulli, T., A.E. Simor and D.E. Low.** 1990. Reproducibility of interpretation of gram-stained vaginal smears for the diagnosis of bacterial vaginosis. J. Clin. Microbiol. **28**:1506-1508.

167. **Meislin, H.W., S.A. Lerner, M.H. Graves, M.D. McGehee, F.E. Kocka, J.A. Morello and P. Rosen.** 1977. Cutaneous abscesses. Anaerobic and aerobic bacteriology and outpatient management. Ann. Intern. Med. **87**:145-149.

168. **Mitsuoka, T., K. Ohno, Y. Benno, K. Suzuki and K. Namba.** 1976. The fecal flora of man. IV Communication: Comparison of the newly developed method with the old conventional method for analysis of intestinal flora. Zentralbl. Bakteriol. [B] **A234**:219-233.

169. **Montgomery, L., B. Flesher and D. Stahl.** 1988. Transfer of *Bacteroides succinogenes* (Hungate) to *Fibrobacter* gen.nov.as *Fibrobacter succinogenes* comb. nov.and description of *Fibrobacter intestinalis* sp.nov. Int. J. Syst. Bacteriol. **38**:430-435.

170. **Moore, L.V.H., E.P. Cato and W.E.C. Moore.** 1987. Anaerobe Laboratory Manual Update. April, 1987. Published as a Supplement to the VPI Anaerobe Laboratory Manual, 4th ed., 1977, Dept. Anaerobic Microbiol., Virginia Polytechnic Institute and State University, Blacksburg, VA.

171. **Moore, L.V.H., J.L. Johnson and W.E.C. Moore.** 1987. *Selenomonas noxia* sp.nov., *Selenomonas flueggei* sp.nov., *Selenomonas infelix* sp.nov., *Selenomonas dianae* sp.nov., and *Selenomonas artemidis* sp.nov., from the human gingival crevice. Int. J. Syst. Bacteriol. **36**:271-280.

172. **Moore, L.V.H. and W.E.C. Moore.** 1991. Anaerobe Lab Manual Update. February, 1991. Published as a Supplement to the VPI Anaerobe Laboratory Manual, 4th ed., 1977, Dept. Anaerobic Microbiol., Virginia Polytechnic Institute and State University, Blacksburg, VA.

173. **Moore, W.E.C., E.P. Cato and L.V. Holdeman.** 1969. Review: Anaerobic bacteria of the gastrointestinal flora and their occurrence in clinical infections. J. Infect. Dis. **119**:641-649.

174. **Moore, W.E.C., L.V. Holdeman, E.P. Cato, I.J. Good, E.P. Smith, R.R. Ranney and K.G. Palcanis.** 1984. Variation in periodontal floras. Infect. Immun. **46:**720-726.

175. **Moore, W.E.C., L.V. Holdeman, R.M. Smibert, I.J. Good, J.A. Burmeister, K.G. Palcanis and R.R. Ranney.** 1982. Bacteriology of experimental gingivitis in young adult humans. Infect. Immun. **38:**651-667.

176. **Morris, G.N., J. Winter, E.P. Cato, A.E. Ritchie and V.D. Bokkenheuser.** 1985. *Clostridium scindens* sp.nov., a human intestinal bacterium with desmolytic activity on corticoids. Int. J. Syst. Bacteriol. **35:**478-481.

177. **Müller, A.J.R.** 1966. Microbiological examination of root canals and periapical tissues of human teeth. Odontologisk Tidskrift. Scand. Dent. J. **74:**1-380.

178. **Mulligan, M.E., D.M. Citron, B.T. McNamara and S.M. Finegold.** 1982. Impact of cefoperazone therapy on fecal flora. Antimicrob. Agents Chemother. **22:**226-230.

179. **Murray, P.R.** 1978. Growth of clinical isolates of anaerobic bacteria on agar media: effects of media composition, storage conditions, and reduction under anaerobic conditions. J. Clin. Microbiol. 8:708-714.

180. **Murray, P.R. and J.E. Sondag. 1978.** Evaluation of routine subcultures of macro-scopically negative blood cultures for detection of anaerobes. J. Clin. Microbiol. **8:**427-430.

181. **Murray, P.R., C.J. Weber and A.C. Niles.** 1985. Comparative evaluation of three identification systems for anaerobes. J. Clin. Microbiol. **22:**52-55.

182. **National Committee for Clinical Laboratory Standards,** 1990. Quality assur-ance for commercially prepared microbiological culture media, Vol. 10, No. 14. Approved Standard, NCCLS Publication M22-A, NCCLS, Villanova, PA.

183. **National Committee for Clinical Laboratory Standards,** 1990. Methods for Antimicrobial Susceptibility Testing of Anaerobic Bacteria—Second Edition, Vol. 9, No. 10. Approved Standard, NCCLS Publication M11-A2, NCCLS, Villanova, PA.

184. **Newman, M.G. and T.N. Sims.** 1979. The predominant cultivable microbiota of the periodontal abscess. J. Periodontol. **50:**350-354.

185. **Nichols, R.L. and J.W. Smith.** 1975. Intragastric microbial colonization in common disease states of the stomach and duodenum. Ann. Surg. **182:**557-561.

186. **Okuda, K., T. Kato, J. Shiozu, I. Takazoe and T. Nakamura.** 1985. *Bacteroides heparinolyticus* sp.nov. isolated from humans with periodontitis. Int. J. Syst. Bacteriol. **35:**438-442.

187. **Paisley, J.W., J.E. Rosenblatt, M. Hall and J.A. Washington, II.** 1978. Evalu-ation of a routine anaerobic subculture of blood cultures for detection of anaerobic bacteremia. J. Clin. Microbiol. **8:**764-776.

188. **Pankuch, G.A. and P.C. Appelbaum.** 1986. Agar medium for gas-liquid chroma-tography of anaerobes. Am. J. Clin. Pathol. **85:**82-86.

189. **Parker, R.T. and C.P. Jones.** 1966. Anaerobic pelvic infections and developments in hyperbaric oxygen therapy. Am. J. Obstet. Gynecol. **96:**645-659.

190. **Peach, S. and L. Hayek.** 1974. The isolation of anaerobic bacteria from wound swabs. J. Clin. Pathol. **27:**578-582.

191. **Perry, L.D., J.H. Brinser and H. Kolodner.** 1982. Anaerobic corneal ulcers. Ophthalmology **89:**636-642.

192. **Rosebury, T.** 1962. Microorganisms Indigenous to Man, McGraw-Hill, New York.

193. **Rosenblatt, J.E., A. Fallon and S.M. Finegold.** 1973. Comparison of methods for isolation of anaerobic bacteria from clinical specimens. Appl. Microbiol. **25:**77-85.

194. **Rotheram, E.B., Jr. and S.F. Schick.** 1969. Nonclostridial anaerobic bacteria in septic abortion. Am. J. Med. **46:**80-89.

195. **Sabbaj, J., V.L. Sutter and S.M. Finegold.** 1972. Anaerobic pyogenic liver abscess. Ann. Intern. Med. **77:**629-638.

196. **Salyers, A.A., J.R. Vercellotti, S.E.H. West and T.D. Wilkins.** 1977. Fermentation of mucin and plant polysaccharides by strains of *Bacteroides* from the human colon. Appl. Environ. Microbiol. **33:**319-322.

197. **Sapico, F.L., H.N. Canawati, J.L. Witte, J.Z. Montgomerie, F.W. Wagner, Jr. and A.N. Bessman.** 1980. Quantitative aerobic and anaerobic bacteriology of infected diabetic feet. J. Clin. Microbiol. **12:**413-420.

198. **Schreckenberger, P.C., D.M. Celig and W.M. Janda.** 1988. Clinical evaluation of the Vitek ANI card for identification of anaerobic bacteria. J. Clin. Microbiol. **26:**225-230.

199. **Schreiner, A.** 1979. Anaerobic pulmonary infections. Scand. J. Infect. Dis. Suppl. 77-79.

200. **Schwan, O.** 1979. Biochemical, enzymatic, and serological differentiation of *Peptococcus indolicus* (Christiansen) Sorensen from *Peptococcus asaccharolyticus* (Distaso) Douglas. J. Clin. Microbiol. **9:**157-162.

201. **Shah, H.N. and D.M. Collins.** 1990. *Prevotella,* a new genus to include *Bacteroides melaninogenicus* and related species formerly classified in the genus *Bacteroides.* Int. J. Syst. Bacteriol. **40:**205-208.

202. **Shah, H.N. and M.D. Collins.** 1982. Reclassification of *Bacteroides multiacidus* (Mitsuoka, Terada, Watanabe, and Uchida) in a new genus *Mitsuokella,* as *Mitsuokella multiacidus* comb.nov. Zentralbl. Bakteriol. Parasitenkd. Infektionskr. Hyg. **3:**491-494.

203. **Shah, H.N. and M.D. Collins.** 1982. Reclassification of *Bacteroides hypermegas* (Harrison and Hansen) in a new genus *Megamonas,* as *Megamonas hypermegas* comb.nov. Zentralbl. Bakteriol. Parasitenkd. Infektionskr. Hyg. **3:**394-398.

204. **Shah, H.N. and M.D. Collins.** 1983. Validation of the publication of new names and new combinations previously effectively published outside the IJSB. List No.10. Int. J. Syst. Bacteriol. **33:**438-440.

205. **Shah, H.N. and M.D. Collins.** 1986. Reclassification of *Bacteroides furcosus* Veillon and Zuber (Hauduroy, Ehringer, Urbain, Guillot and Magrou) in a new genus *Anaerorhabdus,* as *Anaerorhabdus furcosus* comb. nov. Syst. Appl. Microbiol. **8:**86-88.

206. **Shah, H.N. and M.D. Collins.** 1986. Validation of the publication of new names and new combinations previously effectively published outside the IJSB. List No.22. Int. J. Syst. Bacteriol. **36:**573-576.

207. **Shah, H.N. and M.D. Collins.** 1988. Proposal for reclassification of *Bacteroides asaccharolyticus, Bacteroides gingivalis,* and *Bacteroides endodontalis* in a new genus, *Porphyromonas.* Int. J. Syst. Bacteriol. **38:**128-131.

208. **Shah, H.N., M.D. Collins, J. Watabe and T. Mitsuoka.** 1985. *Bacteroides oulorum* sp.nov., a nonpigmented saccharolytic species from the oral cavity. Int. J. Syst. Bacteriol. **35:**193-197.

209. **Sheppard, A., C. Cammarata and D.H. Martin.** 1990. Comparison of different medium bases for the semiquantitative isolation of anaerobes from vaginal secretions. J. Clin. Microbiol. **28:**455-457.

210. **Shimada, K., T. Inamatsu and M. Yamashiro.** 1977. Anaerobic bacteria in biliary disease in elderly patients. J. Infect. Dis. **135:**850-854.

211. **Shinjo, T., K. Hiraiwa and S. Miyazato.** 1990. Recognition of biovar C of *Fusobacterium necrophorum* (Flügge) Moore and Holdeman as *Fusobacterium pseudonecrophorum* sp. nov., nom. rev. (ex Prévot 1940). Int. J. Syst. Bacteriol. **40:**71-73.

212. **Slots, J.** 1987. Detection of colonies of *Bacteroides gingivalis* by a rapid fluorescence assay for trypsin-like activity. Oral Microbiol. Immunol. **2:**139-141.

213. **Slots, J., T.V. Potts and P.A. Mashimo.** 1983. *Fusobacterium periodonticum,* a new species from the human oral cavity. J. Dent. Res. **62:**960-963.

214. **Slots, J., T.V. Potts and P.A. Mashimo.** 1984. Validation of the publication of new names and new combinations previously effectively published outside the IJSB. List No.14. Int. J. Syst. Bacteriol. **34:**270-271.

215. **Slots, J. and H.S. Reynolds.** 1982. Longwave UV light fluorescence for identification of black-pigmented *Bacteroides* spp. J. Clin. Microbiol. **16:**1148-1151.

216. **Smith, H.J. and H.B. Moore.** 1988. Isolation of *Mobiluncus* species from clinical specimens by using cold enrichment and selective media. J. Clin. Microbiol. **26:**1134-1137.

217. **Smith, J.W., P.M. Southern, Jr. and J.D. Lehmann.** 1970. Bacteremia in septic abortion: complications and treatment. Obstet. Gynecol. **35:**704-708.

218. **Smith, R.F., R.R. Rogers and C.L. Bettge.** 1972. Inhibition of the indole test reaction by sodium nitrite. Appl. Microbiol. **23:**423-424.

219. **Sondag, J.E., M. Ali and P.R. Murray.** 1979. Relative recovery of anaerobes on different isolation media. J. Clin. Microbiol. **10:**756-757.

220. **Spiegel, C.A., R. Amsel, D. Eschenbach, F.D. Schoenknecht and K. Holmes.** 1980. Anaerobic bacteria in non-specific vaginitis. N. Engl. J. Med. **303:**601-607.

221. **Spiegel, C.A. and K.J. Krueger.** 1986. A selective agar for isolation of *Mobiluncus* from vaginal fluid. Abstr. Annu. Meeting, Am. Soc. Microbiol. C-334, p. 383.

222. **Stackebrandt, E. and H. Hippe.** 1986. Transfer of *Bacteroides amylophilus* to a new genus *Ruminobacter* gen.nov., nom.rev. as *Ruminobacter amylophilus* comb. nov. Syst. Appl. Microbiol. **8:**204-207.

223. **Stackebrandt, E. and H. Hippe.** 1987. Validation of the publication of new names and new combinations previously effectively published outside the IJSB. List. No.23. Int. J. Syst. Bacteriol. **37:**179-180.

224. **Stenson, M.J., D.T. Lee, J.E. Rosenblatt and J.M. Contezac.** 1986. Evaluation of the AnIdent system for the identification of anaerobic bacteria. Diagn. Microbiol. Infect. Dis. **5:**9-15.

225. **Strong, C.A., M. McTeague, P. Summanen, and E.J. Baron.** 1992. Comparison of recovery of anaerobic bacteria using the Anoxomat, anaerobic chamber, and GasPak jar systems. Abstr. Gen. Meeting, Am. Soc. Microbiol. C-359, p. 480.

226. **Summanen, P. and H. Jousimies-Somer.** 1988. Comparative evaluation of RapID ANA and API-20A for identification of anaerobic bacteria. Eur. J. Clin. Microbiol. Infect. Dis. **7:**771-775.

227. **Sundqvist, G. K.** 1976. Bacteriologic studies of necrotic dental pulps. Doctoral thesis, University of Umea, Umea, Sweden.

228. **Sutter, V.L.** 1984. Anaerobes as normal oral flora. Rev. Infect. Dis. **6:**S62-S66.

229. **Sutter, V.L. and W.T. Carter.** 1972. Evaluation of media and reagents for indole-spot tests in anaerobic bacteriology. Am. J. Clin. Pathol. **58:**335-338.

230. **Sutter, V.L. and S.M. Finegold.** 1971. Antibiotic disc susceptibility tests for rapid presumptive identification of gram-negative anaerobic bacilli. Appl. Microbiol. **21:**13-20.

231. **Sutter, V.L., P.T. Sugihara and S.M. Finegold.** 1971. Rifampin-blood-agar as a selective medium for the isolation of certain anaerobic bacilli. Appl. Microbiol. **22:**777-780.

232. **Swartz, M.N.** 1989. Central nervous system infections, p. 156-212. In S.M. Finegold and W.L. George (eds.), Anaerobic Infections in Humans, Academic Press, San Diego.

233. **Swenson, R.M., B. Lorber, T.C. Michaelson and E.H. Spaulding.** 1974. The bacteriology of intra-abdominal infections. Arch. Surg. **109:**398-399.

234. **Swenson, R.M., T.C. Michaelson, M.J. Daly and E.H. Spaulding.** 1973. Anaerobic bacterial infections of the female genital tract. Obstet. Gynecol. **42:**538-541.

235. **Takayuki, E., L. Shu-Lin, H. Yasuhiro and Y. Eiko.** 1990. *Peptostreptococcus hydrogenalis* sp.nov. from human fecal and vaginal flora. Int. J. Syst. Bacteriol. **40:**305-306.

236. **Tanner, A.C.R., S. Badger, C.-H. Lai, M.A. Listgarten, R.A. Visconti and S.S. Socransky.** 1981. *Wolinella* gen.nov., *Wolinella succinogenes* (*Vibrio succinogenes* Wolin *et al.*) comb.nov., and description of *Bacteroides gracilis* sp.nov., *Wolinella recta* sp.nov., *Campylobacter concisus* sp.nov., and *Eikenella corrodens* from humans with periodontal disease. Int. J. Syst. Bacteriol. **31:**432-445.

237. **Tanner, A.C.R., M.A. Listgarten, J.L. Ebersole and M.N. Strzempko.** 1986. *Bacteroides forsythus* sp.nov., a slow-growing, fusiform *Bacteroides* sp. from the human oral cavity. Int. J. Syst. Bacteriol. **36:**213-221.

238. **Tanner, A.C.R., M.N. Strzempko, C.A. Belsky and G.A. McKinley.** 1985. API ZYM and API AnIdent reactions of fastidious oral gram-negative species. J. Clin. Microbiol. **22:**333-335.

239. **Thadepalli, H., S.L. Gorbach and L. Keith.** 1973. Anaerobic infections of the female genital tract: Bacteriologic and therapeutic aspects. Am. J. Obstet. Gynecol. **117:**1034-1040.

240. **Thorpe, T.C., M.L. Wilson, J.E. Turner, J.L. DiGuiseppi, M. Willert, S. Mirrett and L.B. Reller.** 1990. BacT/Alert: an automated colorimetric microbial detection system. J. Clin. Microbiol. **28:**1608-1612.

241. **Totten, P.A., R. Amsel, J. Hale, P. Piot and K.K. Holmes.** 1982. Selective differential human blood bilayer media for isolation of *Gardnerella vaginalis*. J. Clin. Microbiol. **15:**141-147.

242. **Turton, L.J., D.B. Drucker and L.A. Ganguli.** 1982. Effect of incubation time, and calcium carbonate and glucose in the growth medium, upon the fermentation end-product profile of *Clostridium difficile*. Microbios **35:**7-16.

243. **Turton, L.J., D.B. Drucker and L.A. Ganguli.** 1982. Effect of glucose concentration in the growth medium upon neutral and acidic fermentation end-products of *Clostridium bifermentans, Clostridium sporogenes,* and *Peptostreptococcus anaerobius.* J. Med. Microbiol. **16:**61-67.

244. **Turton, L.J., D.B. Drucker, V.F. Hillier and L.A. Ganguli.** 1983. Effect of eight growth media upon fermentation profiles of ten anaerobic bacteria. J. Appl. Bacteriol. **54:**295-304.

245. **Van Winkelhoff, A.J., M. Clement and J. de Graaff.** 1988. Rapid characterization of oral and nonoral pigmented *Bacteroides* species with the ATB Anaerobes ID system. J. Clin. Microbiol. **26:**1063-1065.

246. **Van Winkelhoff, A.J., T.J. van Steenbergen, N. Kippuw and J. de Graaff.** 1985. Further characterization of *Bacteroides endodontalis,* an asaccharolytic black-pigmented *Bacteroides* species from the oral cavity. J. Clin. Microbiol. **22:**75-79.

247. **Vandamme, P., E. Falsen, R. Rossau, B. Hoste, P. Segers, R. Tytgat and J. De Ley.** 1991. Revision of *Campylobacter, Helicobacter,* and *Wolinella* taxonomy: emendation of generic descriptions and proposal of *Arcobacter* gen.nov. Int. J. Syst. Bacteriol. **41:**88-103.

248. **Washington, J.A., II.** 1978. Conventional approaches to blood culture. p. 41-88. In J.A. Washington II (ed.), The detection of septicemia. CRC Press, West Palm Beach, FL.

249. **Watabe, J., Y. Benno and T. Mitsuoka.** 1983. Taxonomic study of *Bacteroides oralis* and related organisms and proposal of *Bacteroides veroralis* sp.nov. Int. J. Syst. Bacteriol. **33:**57-64.

250. **Watt, B. and E.P. Jack.** 1977. What are anaerobic cocci? J. Med. Microbiol. **10:**461-468.

251. **Weinberg, L.G., L.L. Smith and A.H. McTighe.** 1983. Rapid identification of the *Bacteroides fragilis* group by bile disk and catalase tests. Lab. Med. **14:**785-788.

252. **Wexler, H.M.** 1991. Anaerobic susceptibility testing: myth, magic, or method? Clin. Microbiol. Rev. **4:**470-484.

253. **Wexler, H.M., E. Molitoris and S.M. Finegold.** 1992. In vitro activities of three of the newer quinolones against anaerobic bacteria. Antimicrob. Agents Chemother. **36:**239-243.

254. **Wexler, H.M., E. Molitoris, F. Jashnian and S.M. Finegold.** 1991. Comparison of spiral gradient and conventional agar dilution for susceptibility testing of anaerobic bacteria. Antimicrob. Agents Chemother. **35:**1196-1202.

255. **Whiley, R.A. and J.M. Hardie.** 1989. DNA-DNA hybridization studies of phenotypic characteristics of strains within the 'Streptococcus milleri group.' J. Gen. Microbiol. **135:**2623-2633.

256. **Whitehead, S.M., R.D. Leach, S.J. Eykyn and I. Phillips.** 1982. The aetiology of perirectal sepsis. Br. J. Surg. **69:**166-168.

257. **Wideman, P.A., D.M. Citronbaum and V.L. Sutter.** 1977. Simple disk test for detection of nitrate reduction by anaerobic bacteria. J. Clin. Microbiol. **5:**315-319.

258. **Wideman, P.A., V.L. Vargo, D. Citronbaum and S.M. Finegold.** 1976. Evaluation of the sodium polyanethol sulfonate disk test for the identification of *Peptostreptococcus anaerobius.* J. Clin. Microbiol. **4:**330-333.

259. **Wilkins, T.D., D.L. Wagner, B.J. Veltri, Jr. and E.M. Gregory.** 1978. Factors affecting production of catalase by *Bacteroides.* J. Clin. Microbiol. **8:**553-557.

260. **Wilks, M., R.N. Thin and S. Tabaqchali.** 1984. Quantitative bacteriology of the vaginal flora in genital disease. J. Med. Microbiol. **18:**217-231.

261. **Williams, B.L., G.F. McCann and F.D. Schoenknecht.** 1983. Bacteriology of dental abscesses of endodontic origin. J. Clin. Microbiol. **18:**770-774.

262. **Wilson, K.H., M.J. Kennedy and F.R. Fekety.** 1982. Use of sodium taurocholate to enhance spore recovery on a medium selective for *Clostridium difficile.* J. Clin. Microbiol. **15:**443-446.

263. **Wilson, M.L., M.P. Weinstein, L.G. Reimer, S. Mirrett, and L.B. Reller.** 1991. Controlled comparison of BacT/Alert and BACTEC NR 660/730 blood culture systems. Abstr. Gen. Meeting, Am. Soc. Microbiol. C-365, p. 403.

264. **Wren, B.W., C.L. Clayton, N.B. Castledine and S. Tabaqchali.** 1990. Identification of toxigenic *Clostridium difficile* strains by using a toxin A gene-specific probe. J. Clin. Microbiol. **28:**1808-1812.

265. **Wu, T.C. and S.M. Gersch.** 1986. Evaluation of a commercial kit for the routine detection of *Clostridium difficile* cytotoxin by tissue culture. J. Clin. Microbiol. **23:**792-793.

266. **Wüst, J.** 1977. Presumptive diagnosis of anaerobic bacteremia by gas-liquid chromatography of blood cultures. J. Clin. Microbiol. **6:**586-590.

267. **Wyss, C.** 1989. Dependence of proliferation of *Bacteroides forsythus* on exogenous *n*-acetylmuramic acid. Infect. Immun. **57:**1757-1759.

268. **Zylber, L.J. and H.V. Jordan.** 1982. Development of a selective medium for detection and enumeration of *Actinomyces viscosus* and *Actinomyces naeslundii* in dental plaque. J. Clin. Microbiol. **15:**253-259.

INDEX

Page numbers followed by *t* refer to tables, and page numbers followed by *f* refer to figures.